Real Life Problems in Maths

Garry Norman

Edward Arnold

© Garry Norman 1986

First published in Great Britain 1986 by
Edward Arnold (Publishers) Ltd, 41 Bedford Square, London WC1B 3DQ

Edward Arnold (Australia) Pty Ltd, 80 Waverley Road, Caulfield East,
Victoria 3145, Australia

Reprinted 1987

British Cataloguing in Publication Data

Norman, Garry
    Everyday problems in maths.
    1. Mathematics—Problems, exercises, etc.
    I. Title
    510'.76   QA43

    ISBN 0-7131-8418-3

Text set in 11/13 Century
by D.P. Press Ltd, Sevenoaks
Printed and bound by
Thomson Litho Ltd, East Kilbride

Preface

This book provides pupils with first-hand experience of using mathematics in situations that they will encounter outside and beyond the classroom. To this end, I have made extensive use of real material, and placed it in the context of situations with which pupils will be familiar. Particular emphasis is placed throughout on problem solving, information handling and understanding.

Each chapter is self-contained, and does not require any preceding chapter to have been completed. Teachers may therefore select the chapters in any order.

The symbol ◆ denotes that a photocopy master sheet is available which pupils will write on or cut up. All of the photocopy masters used within a chapter consist of real material, such as booking forms, order forms, and grids for drawing on, and are therefore, in a sense, only one step away from the 'real world'. In addition to these, there are extra sheets at the end of each chapter which provide further extension work for the more able pupils in the class. These can be handed out to pupils who have reached the end of the chapter to work through on their own.

In many cases, calculators are needed, and I have assumed that these are freely available in class, as this, again, is likely to reflect the situation outside the classroom.

I would hope that many teachers would extend the work generated by the book to encourage pupils to investigate each topic further, by collecting more real material and answering questions raised by the pupils themselves.

Finally, I am indebted to my wife, Penny, without whose help, advice and perseverance this book could not have been written.

G. Norman, 1986

Acknowledgements

The publishers would like to thank the following for permission to include copyright material:

Bristol Waterworks Company for their bill and leaflet on metering; Bryan Brothers Ltd for their advertisement; Cheltenham Racecourse for their fixtures; Comet Group plc for material from their catalogue; Co-operative Retail Services Ltd for their advertisement; Ford Motor Company Limited for their catalogue; Greenfields Mortgage and Financial Services for their advertisement; Horizon Holidays for their Summer Holidays catalogue; Ladbroke Dragonara Hotel Bristol for their menus and price list; MFI Furniture Centres Ltd for their kitchen planning leaflet; Moss Advertising and Publicity Ltd for their advertisement; and Stoke Lane Motoring Centre Ltd for their advertisement.

Cartoons by Val Saunders.

Contents

		page
1	Titanic Frozen Foods	7
2	A Day at the Races	11
3	The Budget Account	15
4	The Water Meter	20
5	TV and Video	24
6	The Wedding Reception	28
7	Summer Holidays	34
8	Planning a Kitchen	39
9	Cars	44

1 Titanic Frozen Foods

David Hughes has just become a Sales Representative for the Titanic
Frozen Foods Company. He lives in Bristol and so every day he leaves
Bristol, visits other towns and then, at the end of the day, returns to
Bristol.

Before he starts his job he has to work out which towns he will visit on
which days.

Now read David's job specification on page 8, and then look at the map
on page 9.

1 What is the shortest distance from:
   (a) Bristol to Wells
   (b) Bristol to Frome
   (c) Bath to Castle Cary?

2 On one day he is going to visit Bath, Radstock, Frome and Westbury
   and then return to Bristol. Which route covers the minimum
   distance? What is the distance?

3 How much can he claim in expenses for the journey:
   (a) from Bristol to Cheddar and back
   (b) from Bristol to Frome via Bath and Westbury?

Titanic Frozen Foods

Head Office:
Berg House,
Ross Street,
London W1A 2RP

Telephone: 01-234 9876

Job Specification for:

SALES REPRESENTATIVE FOR REGION G5

To visit at least twice a week the retail freezer centres in the
following towns:-

Bristol, Bath, Westbury, Frome, Radstock, Cheddar, Wells,
Shepton Mallet, Street and Castle Cary.

At each freezer centre, the Sales Representative must encourage the
owners to order as much of our produce as possible. He/she will take
an order from the owners and at the end of the day telephone the orders
through to Head Office so that they can be sent out during the next
working day from our freezer store in Bristol.

Each representative is required to submit a detailed plan of his or her
routes each week to his local manager. Please ensure that your route
plan keeps the number of miles covered to a minimum. When planning a
weekly route, remember that representatives will work for about 40 hours
per week, and that they will cover about 30 miles per hour when
travelling. Each visit to a town will take about $1\frac{1}{2}$ hours. Remember
also that stores are open from 9 a.m. to 6 p.m. from Monday to Saturday.

A car (Ford Sierra 1.6L) is provided and travelling expenses can be re-
claimed at the end of each month.

The salary is £6612 per annum plus 2% commission on all sales over £200
per week. In addition to public holidays, a Sales Representative may
take 4 weeks annual paid leave.

Titanic Frozen Foods

Head Office:
Berg House,
Ross Street,
London W1A 2RP

Telephone: 01-234 9876

INFORMATION SHEET

REGION G5

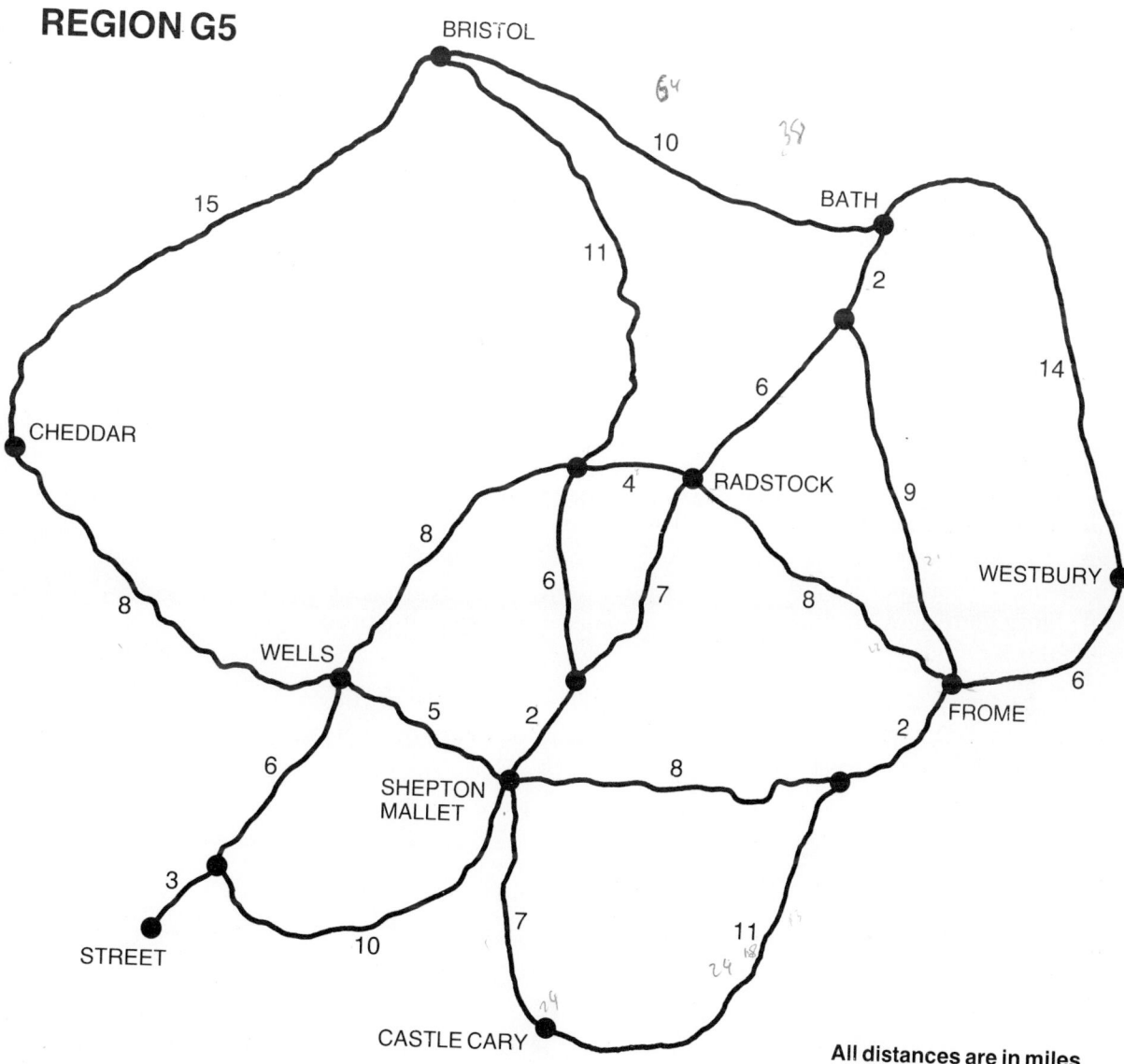

All distances are in miles

TRAVELLING EXPENSES

engine size	per mile
0 — 999 c.c.	15.3p
1000 — 1499 c.c.	17.4p
1500 — 1999 c.c.	20.3p
2000 and over	23.4p

9

To work out the time that a journey takes:

Find the total distance, divide by 30, (because he travels at 30 m.p.h.) **and then multiply by 60 to find how many minutes.**

This is the same as just multiplying the number of miles by 2. For example:

From Bristol to Bath and back is 20 miles, so it would take David
$$2 \times 20 = 40 \text{ minutes.}$$
When he arrives at a store he stays for 1½ hours, so his total time is
$$1\tfrac{1}{2} \text{ hrs} + 40 \text{ minutes} = 2 \text{ hrs } 10 \text{ mins.}$$

4 Find the total time for a journey from Bristol to a store in Cheddar and back.

5 Find the total time for a journey involving stops at stores in Wells, Street and Shepton Mallet from Bristol and back. (Use the shortest route)

6 One day David plans to leave his home at 8.45 a.m., travel to Bath, Westbury, Frome and Radstock and then return home. Using the shortest route and including stopping times, work out the time he gets home. Could he leave any earlier than 8.45 a.m.?

Extension Work

7 (a) Collect a Weekly Route Plan Sheet. Look again at the map and job specification and remember that David must visit each town at least twice per week. He hopes that he does not have to work on Saturdays. Work out David's weekly route plan.
(b) Using your plan, find where and when David will have his lunch each day.

8 The delivery van needs to stop in each town for ½ hr to deliver the orders which David has won. Work out routes for delivery to each town, if each town needs two deliveries per week.

2 A Day at the Races

A group of people want to go to Cheltenham Racecourse to see the races.
Look at the list of 1985 fixtures below:
The three days of 12, 13 and 14 March are known as the National Hunt
Festival (NH).

1 How many Standard Days are there in 1985?

2 How many race meetings at Cheltenham take place on Saturdays?

3 When is the race for the Mackeson Gold Cup run?

4 Arkle is the name of a famous horse. On what day is there a race
named after him?

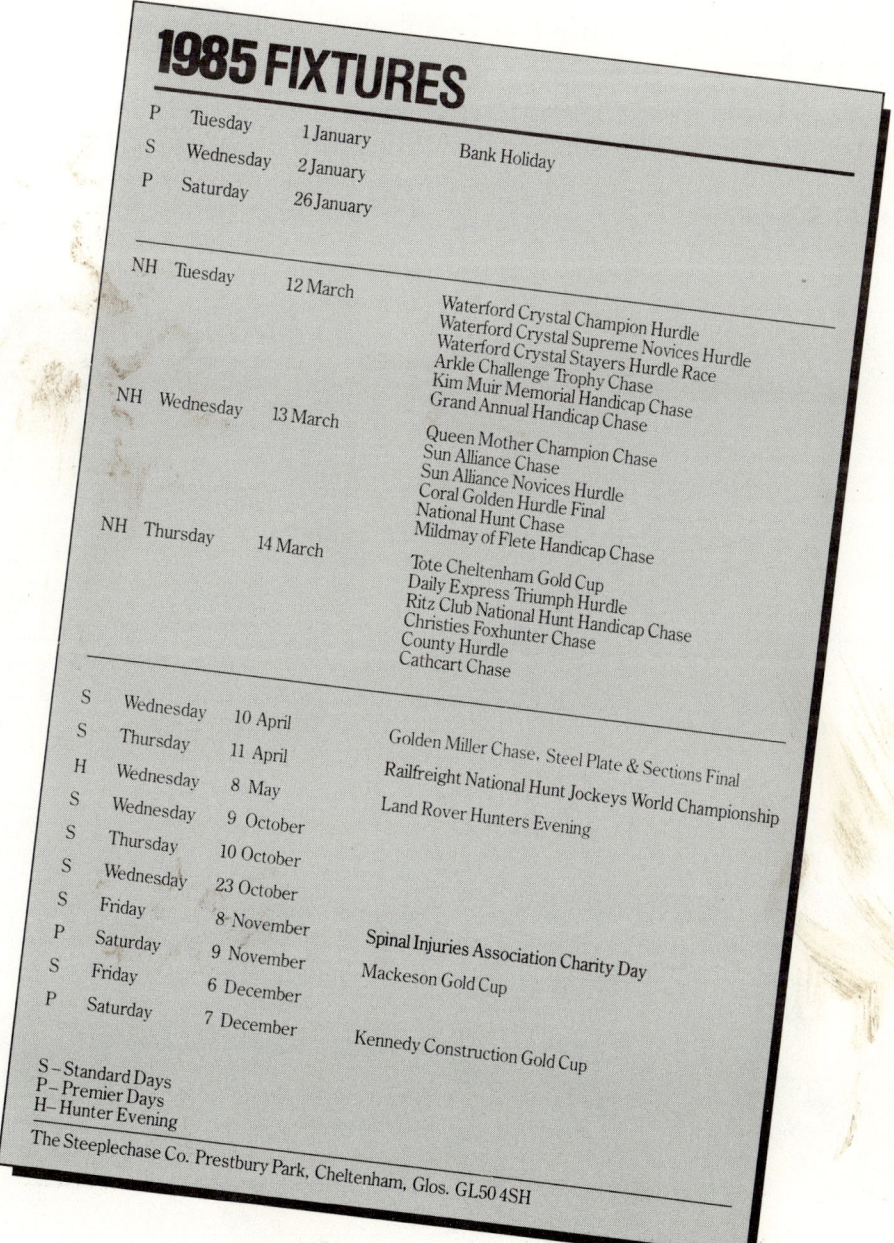

1985 FIXTURES

P	Tuesday	1 January	
S	Wednesday	2 January	Bank Holiday
P	Saturday	26 January	
NH	Tuesday	12 March	Waterford Crystal Champion Hurdle Waterford Crystal Supreme Novices Hurdle Waterford Crystal Stayers Hurdle Race Arkle Challenge Trophy Chase Kim Muir Memorial Handicap Chase Grand Annual Handicap Chase
NH	Wednesday	13 March	Queen Mother Champion Chase Sun Alliance Chase Sun Alliance Novices Hurdle Coral Golden Hurdle Final National Hunt Chase Mildmay of Flete Handicap Chase
NH	Thursday	14 March	Tote Cheltenham Gold Cup Daily Express Triumph Hurdle Ritz Club National Hunt Handicap Chase Christies Foxhunter Chase County Hurdle Cathcart Chase
S	Wednesday	10 April	
S	Thursday	11 April	Golden Miller Chase, Steel Plate & Sections Final
H	Wednesday	8 May	Railfreight National Hunt Jockeys World Championship
S	Wednesday	9 October	Land Rover Hunters Evening
S	Thursday	10 October	
S	Wednesday	23 October	
S	Friday	8 November	
P	Saturday	9 November	Spinal Injuries Association Charity Day Mackeson Gold Cup
S	Friday	6 December	
P	Saturday	7 December	Kennedy Construction Gold Cup

S – Standard Days
P – Premier Days
H– Hunter Evening

The Steeplechase Co. Prestbury Park, Cheltenham, Glos. GL50 4SH

When people go to the racecourse they can go into three different enclosures, called the Club, Tattersalls and Course.

Look at the admission rates shown below:

5 Which of the three enclosures is always the cheapest?

6 What is the entrance fee for Tattersalls on 9 November?

7 How much does it cost to go into the Club on 9 October?

8 On which day of the year can you pay the most to go in, and which enclosure would this be?

9 What would be the total cost for one person to park their car and go into Tattersalls on the day of the Sun Alliance Chase?

10 Four friends want to see the Waterford Crystal Champion Hurdle. They will need to park their car and they want to go into Tattersalls.
   (a) What will the total cost be for all four people?
   (b) How much will each pay if the costs are shared equally?

11 Which is the cheaper, to go into the Club on 26 January or into Tattersalls on 13 March?

12 Mr. and Mrs. Marks and their 13 year-old daughter wish to park their car and go into Tattersalls. How much will it cost them if they go on:
   (a) 10 October?
   (b) 14 March?

ADMISSION RATES

		Club Daily	Tattersalls	Course	Car Park
STANDARD Days		£7	£5	£2	Free
PREMIER Days		£9	£6	£2	Free
HUNTER Evening		£6	£6	£2	Free

NATIONAL HUNT FESTIVAL	Festival Grandstand Seats	Club Daily	Tattersalls	Course	Car Park
Tuesday	£7	£18	£10	£4	£3
Wednesday	£7	£18	£10	£4	£3
Thursday	£10	£20	£13	£5	£3

3 Day non-refundable Advance Purchase Club Badge £50

Badges and tickets are available in advance from 6th February. Apply to:
The Steeplechase Company, Prestbury Park, Cheltenham, Glos. GL50 4SH
The Racing Information Bureau, Winkfield Road, Ascot, Berks. SL5 7HX
The Jockey Club, 42 Portman Square, London W1H 0EN (Personal Application Only)
Keith Prowse Sports Dept., 24 Store Street, London WC1E 7BA (01-631-3380)

CHILDREN under 16 years accompanied by an adult are admitted FREE to all enclosures EXCEPT into the Club at the National Hunt Festival, when full rates will apply.

Prices and other information in this publication are subject to alteration.

If you become a Full Member, you can get into the Club free on any race day and park your car for no charge.

Fred wants to go to three days of the National Hunt Festival, the Mackeson Gold Cup and the Kennedy Construction Gold Cup. He needs to park his car and he goes into the Club.

Look at the Annual Membership details:

13 How much must Fred pay to become a member?

14 If Fred did not become a member, how much would it cost him to attend the races on the five days mentioned?

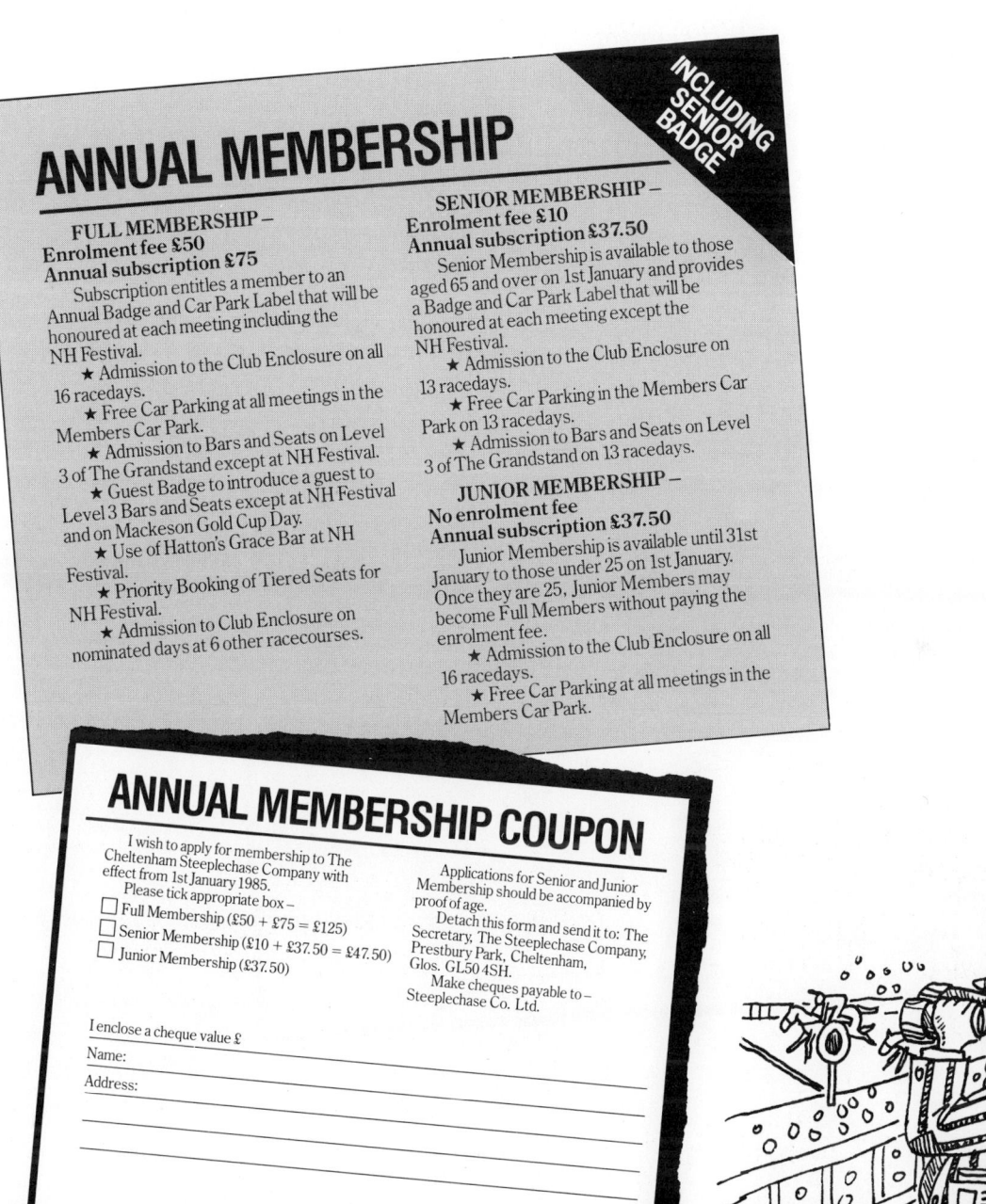

ANNUAL MEMBERSHIP

INCLUDING SENIOR BADGE

FULL MEMBERSHIP –
Enrolment fee £50
Annual subscription £75

Subscription entitles a member to an Annual Badge and Car Park Label that will be honoured at each meeting including the NH Festival.
★ Admission to the Club Enclosure on all 16 racedays.
★ Free Car Parking at all meetings in the Members Car Park.
★ Admission to Bars and Seats on Level 3 of The Grandstand except at NH Festival.
★ Guest Badge to introduce a guest to Level 3 Bars and Seats except at NH Festival and on Mackeson Gold Cup Day.
★ Use of Hatton's Grace Bar at NH Festival.
★ Priority Booking of Tiered Seats for NH Festival.
★ Admission to Club Enclosure on nominated days at 6 other racecourses.

SENIOR MEMBERSHIP –
Enrolment fee £10
Annual subscription £37.50

Senior Membership is available to those aged 65 and over on 1st January and provides a Badge and Car Park Label that will be honoured at each meeting except the NH Festival.
★ Admission to the Club Enclosure on 13 racedays.
★ Free Car Parking in the Members Car Park on 13 racedays.
★ Admission to Bars and Seats on Level 3 of The Grandstand on 13 racedays.

JUNIOR MEMBERSHIP –
No enrolment fee
Annual subscription £37.50

Junior Membership is available until 31st January to those under 25 on 1st January. Once they are 25, Junior Members may become Full Members without paying the enrolment fee.
★ Admission to the Club Enclosure on all 16 racedays.
★ Free Car Parking at all meetings in the Members Car Park.

ANNUAL MEMBERSHIP COUPON

I wish to apply for membership to The Cheltenham Steeplechase Company with effect from 1st January 1985.
Please tick appropriate box –
☐ Full Membership (£50 + £75 = £125)
☐ Senior Membership (£10 + £37.50 = £47.50)
☐ Junior Membership (£37.50)

Applications for Senior and Junior Membership should be accompanied by proof of age.
Detach this form and send it to: The Secretary, The Steeplechase Company, Prestbury Park, Cheltenham, Glos. GL50 4SH.
Make cheques payable to – Steeplechase Co. Ltd.

I enclose a cheque value £
Name:
Address:

Signature:
Date:

Extension Work

Twenty-seven adults and three teenagers aged 14 are arranging a coach trip to Cheltenham Races. They will go into Tattersalls and the coach will cost £75.

15 What will be the total cost for the party (including the coach, parking and entrance fee) if they make a Party Booking for:
(a) 23 October
(b) 26 January
(c) the Tote Cheltenham Gold Cup?

16 For question 15(b) calculate a **fair** share of the cost for each adult and each of the teenagers.

17 Collect a Party Booking Sheet and complete **one** of the booking coupons for this party going into Tattersalls for the Land Rover Hunters Evening.

18 Complete the other booking coupon on the Party Booking Sheet for your family to go to a meeting. On the back of the Sheet, make a list of the adults in the party, and another of the people paying **less** than the adult charge, with their ages.

PARTY BOOKING

A discount of 20%* is available to all parties of 20 and over, booked and paid for in advance to each meeting into Tattersalls and to the NH Festival into the Course Enclosure.
● Free coach parking, close to the grandstands.
● Free admission for coach drivers and children under 16 years of age.

*SPECIAL RATE
NH Festival WEDNESDAY.
On 13th March – Sun Alliance Day – a special rate of £6.60 is available for Tattersalls.

		TATTERSALLS		COURSE	
		Normal Price	Party Rate	Normal Price	Party Rate
NH FESTIVAL	Tuesday	£10	£8	£4	£3.20
	Wednesday	£10	£6.60	£4	£3.20
	Thursday	£13	£10.40	£5	£4
PREMIER	Days	£6	£4.80		
STANDARD	Days	£5.00	£4.00		
HUNTER	Evening	£6.00	£4.80		

3 The Budget Account

Bad news on the doorstep

Tim and Jane were recently married and they moved into a new house. For three months they were very happy, and then one day through the letter box dropped a gas bill for £74.63, an electricity bill for £58.72 and a reminder that in two weeks time Tim had to pay his car insurance premium of £103.47. With only £50 in the bank, Tim and Jane did not know what to do.

1 How much did Tim and Jane owe on these three bills altogether?

Tim and Jane decided to write to their bank manager to ask for help. In their letter they asked for a loan to cover the bills, and this is the reply that they received:

National Bank plc

56 High Street
Middleton
Wessex BR3 9FG

Telephone:
Middleton (STD 0687) 544378

our ref: ASP/JG/2

4th June 1985

T F Mather Esq.
24 Mill Grove
Middleton BR4 7HK

Dear Mr Mather,

re: Current Account No. 91029800

Thank you for your letter requesting a loan to cover your recent and unexpected bills. While I am quite happy to arrange a loan for you, I feel that it would be better for your future financial planning if you were to open a Budget Account.

I have enclosed a leaflet explaining how to open a Budget Account and how the account works, so that unexpected bills do not cause unnecessary problems. I have also enclosed an application form for you to apply for a Budget Account if you so wish.

I hope that this will solve your present and future problems and I look forward to receiving your completed form.

Yours sincerely,

A S Pound
Manager

Bills get lonely very easily, so they like to arrive together. Bills have an unhappy talent for arriving in batches. Some months are worse than others.

If you're on a regular salary, paying them can be quite difficult. Because, after all, you've got to keep some money in hand to live on.

This is what you do
Make a list of all the household bills you expect to receive during the coming year.

Add together rates, telephone, car expenses, gas and electricity as well as holidays, Christmas and any other annual expenses.

Then divide the total (which must amount to a minimum of £500), by twelve. If you prefer transfers can be made every four weeks, ie total expenses can be divided by thirteen rather than twelve.

Complete and return both copies of the schedule together with the agreement to your branch.
If you do not already have an account with National Bank simply take the forms to the most convenient branch.

This is what we do
The National Bank will arrange to transfer each one of those twelve or thirteen instalments from your Current Account to a Budget Account every month.

Once the first transfer has been made you may write out cheques to pay your bills as they come in. We do ask you to agree not to make a single payment which exceeds one quarter of the total commitment.

No interest is payable
It's not a free service, but it can pay for itself. The only charge we make is a service charge at the rate of £35 to cover the first £500 plus £1 for every £50 or part of £50 after that, based on the total amount of payments made out of the account. Although we expect the account to run mainly in credit you may overdraw subject to the conditions in the agreement form without incurring any additional expense.

Note: Budget Account facilities are available only to customers over 18 years of age and at the discretion of the Branch Manager.

This is how a typical Budget Account would work

Bills	Amounts
	£
General Rates	320
Water Rates	72
Telephone	160
Electricity	170
Gas	300
Fuel (including Oil and Coal)	—
School Fees	—
Life Assurance	90
House and Contents Insurance	80
Car Insurance	80
Car Licence	90
Servicing and Maintenance Agreements	—
Season Ticket	
Television Licence	46
Holidays	270
Annual Subscriptions	35
Clothing	130
Christmas Expenses	140
Stocking the Freezer	150
	2,133
Estimated total bills for the year	
Add: Service charge	68
	2,201
Add: Savings	150
	2,351
Amount of monthly transfer from Current Account	195.92

Budget Account
How to cope with irregular bills on a regular income

Study the leaflet on page 16 about the Budget Account.

2 What would the service charge be for total bills of £500?

3 What is the service charge for bills totalling £750?

4 What would Tim and Jane pay in service charges if their total bills came to £1238?

5 If the service charge was £42, what would be:
   (a) the smallest
   (b) the greatest
   total of bills?

 6 Collect a Budget Account Schedule and complete it for the example given on page 16.

Tim and Jane write down the bills they expect to pay during a whole year as follows:

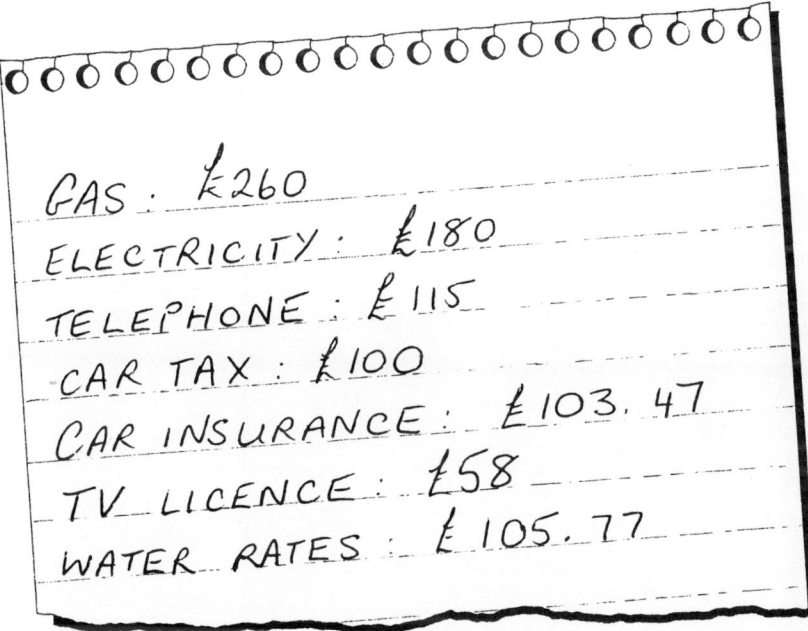

GAS : £260
ELECTRICITY : £180
TELEPHONE : £115
CAR TAX : £100
CAR INSURANCE : £103.47
TV LICENCE : £58
WATER RATES : £105.77

 7 Complete a Budget Account Schedule for Tim and Jane for their expected bills. Do not forget to divide the total by twelve to find the monthly payments.

Extension Work

8 What sort of things could be paid for under the headings 'Season Ticket', 'Annual Subscriptions' and 'Servicing and Maintenance Agreements'?

9 Make up a schedule for your own estimates of bills, but make sure that the final monthly payment is not over £100. You may have to leave out some items like holidays!

How did Tim and Jane manage?

You will have completed a schedule for Tim and Jane in question 7. Now we can see how they managed. Did they plan correctly, or did the bills come to more than they had thought?

 10 Collect a Budget Account Record Sheet. This is for Tim and Jane to keep a record of all the bills that come in so that, at the end of a year, they can see if they have done their sums correctly.

Below is a list of all the bills that Tim and Jane paid during the year. Complete the record sheet for each bill and then add up the amount spent on each category of the account.

Date	Bill	Amount
3 May	Electricity	£37.29
7 June	Gas	£58.14
13 July	Telephone	£35.62
3 August	Electricity	£30.16
6 Sept	Gas	£38.72
8 Oct	Water Rates	£49.27
12 Oct	Telephone	£24.37
1 Nov	Car Insurance	£95.58
1 Nov	Car Licence	£100.00
4 Nov	Electricity	£35.45
8 Dec	Gas	£75.91
1 Jan	TV Licence	£58.00
13 Jan	Telephone	£29.84
4 Feb	Electricity	£42.31
7 March	Gas	£89.73
8 April	Water Rates	£54.89
11 April	Telephone	£28.42

11 Did Tim and Jane end the year (at the end of April) with some money in the account, or did they still owe money? How much?

12 Make a list of things which they **under**estimated, and another list of things which they **over**estimated. For each item in your lists, write down the difference between their estimate and the actual amount.

Extension Work

13 The monthly instalment that Tim and Jane pay into their Budget Account is £80.52, and it is paid on the 2nd day of every month starting on 2 May.

Collect and complete a Bank Balance Sheet to show the balance of Tim and Jane's Budget Account as they make payments **to** the account and pay bills **from** the account. The first few lines are shown below.

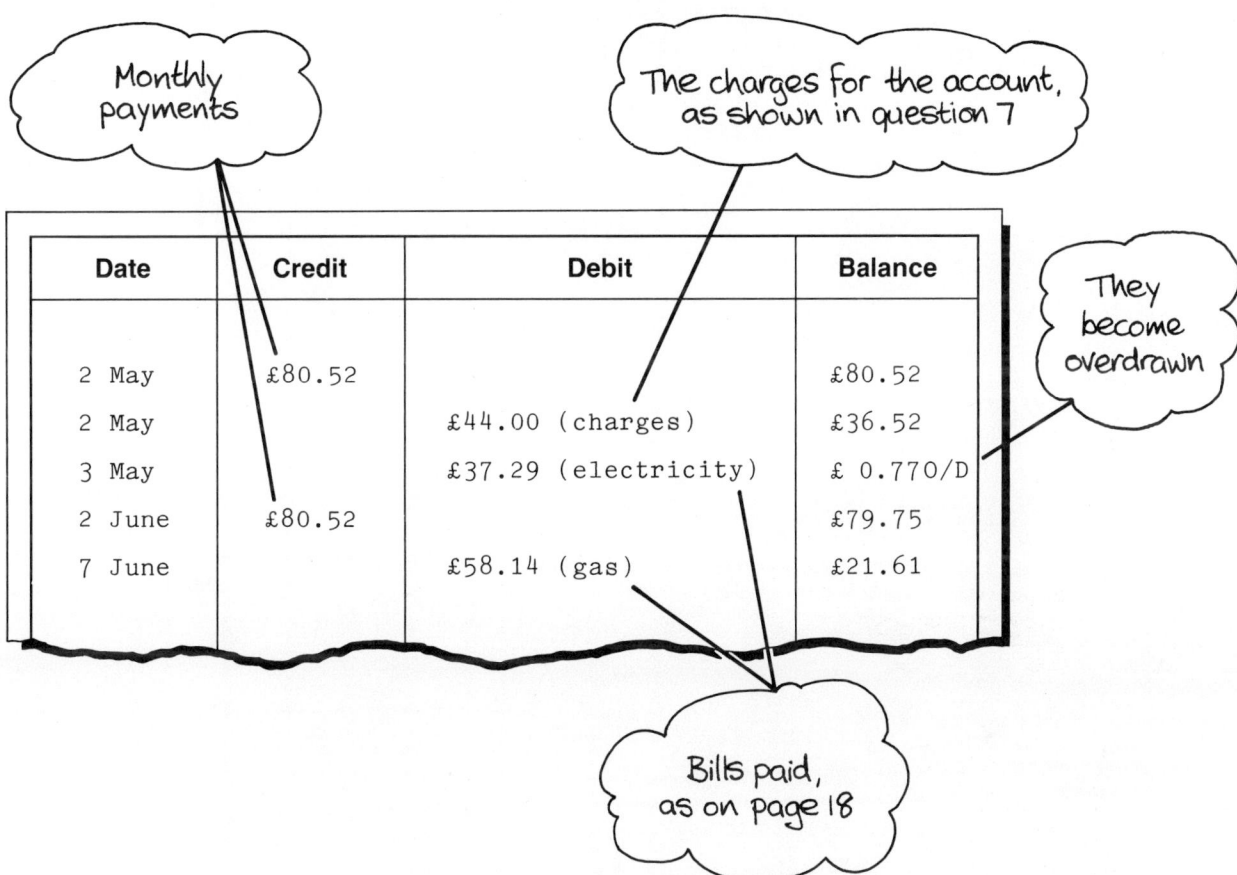

Date	Credit	Debit	Balance
2 May	£80.52		£80.52
2 May		£44.00 (charges)	£36.52
3 May		£37.29 (electricity)	£ 0.77O/D
2 June	£80.52		£79.75
7 June		£58.14 (gas)	£21.61

Monthly payments

The charges for the account, as shown in question 7

They become overdrawn

Bills paid, as on page 18

14 Do you think that having a Budget Account was a good idea for Tim and Jane? Why?

15 Tim and Jane think that next year the TV licence and car tax will stay the same, but that water rates will go up by 5%, gas and electricity by 8%, telephone by 6% and car insurance by 15%.

Look at how much they **actually paid** for these things this year, and calculate how much they will need to pay next year. By drawing up a new schedule find out how much extra they will have to pay into the next year's Budget Account each month to meet the new bills.

4 The Water Meter

Mr. Stern has received a bill from Bristol Waterworks Company for all the water used in his house. Look at the bill below.
The bill is worked out like this:

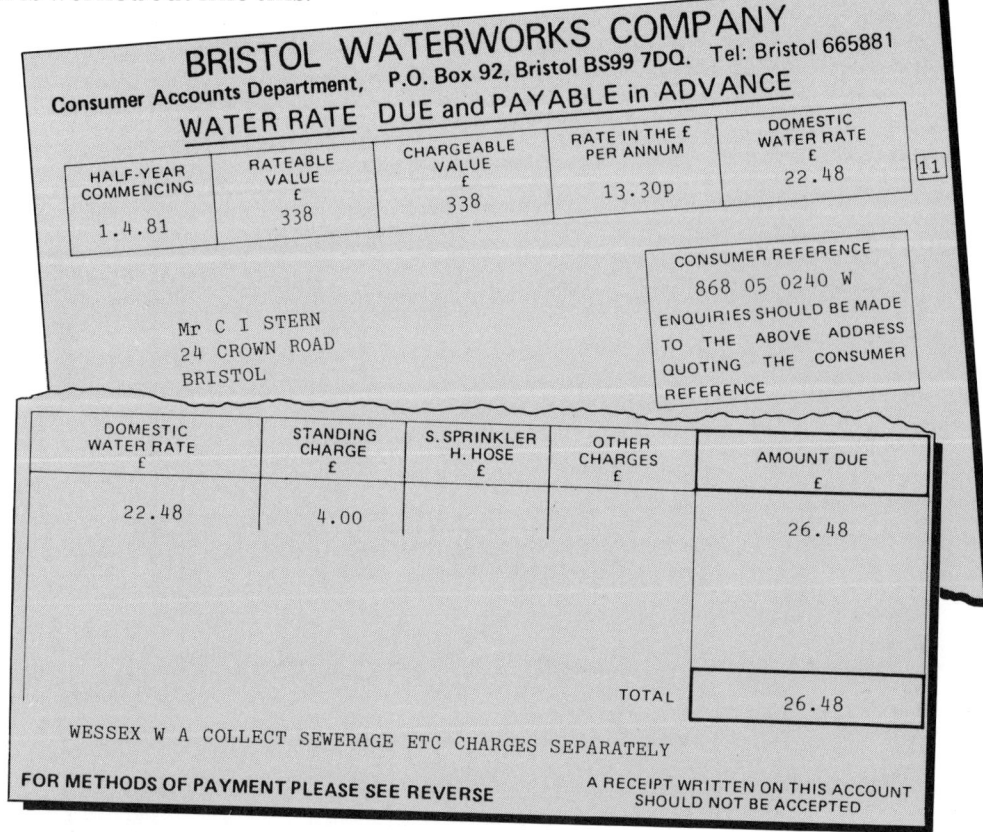

Mr. Stern's house has a **rateable value** (decided by his local council) and the Waterworks Company fix a **rate in the pound**. To calculate Mr. Stern's bill, the rateable value (£338) is multiplied by the rate in the pound (13.30p) and then divided by 2 (for half a year) to give £22.48. Finally, a **standing charge** of £4.00 is added to give a total of £26.48. This is how much Mr. Stern has to pay for **half a year**.

1 (a) How much will Mr. Stern pay for a **whole** year?
   (b) How much would he pay in a whole year if the standing charge was increased to £4.37 for six months?

2 Mr. Holmes lives at 34 Crown Road, and the rateable value of his house is £276. Find out how much Mr. Holmes will pay for half a year with a standing charge of £4.00.

3 These bills were for last year, and this year the charges have gone up. The rate in the pound is now 14.40p and the standing charge is £9.50 for a **whole** year.
   Calculate the new bills for this year for:
   (a) Mr. Stern
   (b) Mr. Holmes.

This year people can choose to pay their water rates in a different way. They can have a meter fitted to their house which measures exactly how much water is used, rather like a gas or electric meter. This means that if you use more water than usual you will pay more, and if you reduce the amount of water that you use you will pay less.

Now read the document below sent to all the people living in Crown Road.

WATER METERING

PLEASE READ THIS LEAFLET CAREFULLY

BRISTOL WATERWORKS COMPANY, PO Box 92, Bridgwater Road, Bristol BS99 7DQ (Tel: Bristol 665881)

You could save money with a water meter if:

* You live alone or have a small household
* You don't use a lot of water
* Your rateable value is high

YOUR CHOICE

From April 1983 all consumers in the Bristol Waterworks area of supply will be given the option to pay for water by means of a metered supply instead of the present rateable value basis.

However, although the rateable value method of charging can produce anomalies, in the majority of cases the cost of water by meter will be roughly equal to the amount paid on the present system.

Our advice is to consider your own consumption and circumstances carefully before applying for a meter. To this end we have included some notes to help you.

DOMESTIC WATER USE

The following statistics show AVERAGE consumptions for certain common domestic activities:

* Drinking, washing up, cooking, personal hygiene (excluding baths and showers)	41 litres per person per day
* Toilet flushing	9 litres per flush
* Bath	91 litres per bath
* Shower	27 litres per person
* Washing machine, top loader	136 litres per cycle
* Washing machine, front loader	100 litres per cycle
* Washing machine, Twin Tub	114 litres per load
* Washing machine, Single Tub	77 litres per load
* Dishwasher	59 litres per cycle
* Hosepipe/Sprinkler	9 litres per minute
* Swimming Pool	27 litres per cubic foot of capacity

Note
Modern water meters register in cubic metres and bills are calculated accordingly. 1 cubic metre = 1000 litres

PATTERN OF USAGE

Not all households consume water in the same way and it may be helpful to decide whether your total use of water is high, average or low. For example:-

Most people who frequently use dishwashers or washing machines tend to be high users.

Others who don't use either, don't use garden watering appliances and don't take more than one bath or shower a week tend to be low users.

If you feel you don't fit into either of the above you should regard yourself as an average user

The following table shows some typical water use figures (in cubic metres per year):

Number of Persons in household	High Use	Average Use	Low Use
1	60	50	40
2	135	90	70
3	180	135	110
4	220	150	135
5	260	170	150
6 & over	280	180	150

THIS YEAR'S CHARGES

Our charges, for water supply only, for the year starting 1st April 1983 are as follows:

1   UNMEASURED TARIFF - 14.4p in the £ based on the rateable value of your property, plus a standing charge of £9.50 a year. In addition, charges for a full year will be made for:

Hand-held hosepipe	£7
Sprinkler/unattended hosepipe	£21
Unmetered field trough	£19

It is illegal to use any of these unless you give advance notice and undertake to pay the charge.

2   MEASURED TARIFF - A standing charge based on the size of meter (e.g. £30 a year for a ½" supply - normal domestic supply) plus a volume charge as measured by the meter of 20.5p per cubic metre. Hosepipes/sprinklers, etc., are not charged additionally.

If you choose a meter then you have to think about how much water you actually use.

4 Which uses most water—two washes in a single-tub washing machine or one wash in a twin-tub machine?

5 How many showers would you have to take before you used as much water as one bath?

6 How many litres are used in one day in a house where the two people living there each have a shower every morning, and use their front-loading washing machine every **other** day?

7 If a garden sprinkler was left on for 25 minutes, how many bathfuls of water would it use?

8 A house has a swimming pool that measures 10 feet wide, 15 feet long and 6 feet deep.
(**a**) How many litres does it take to fill it?
(**b**) If the pool was filled with a hosepipe, how long would it take to fill it?

9 Make an estimate of how much water is used in your home in one week.

10 Pat Watson has a bath every morning and does her washing at the launderette. She does not have a garden as she lives in a flat. Estimate how much water Pat Watson uses in one year in cubic metres.

4A 11 Now collect a Water Metering Decision Sheet. Complete the sheet (in the 'Your Own' column) for Pat Watson. In order to find the total saving or loss, you need to know the rateable value of her flat. This can be found by looking at the top part of her last bill, shown below.

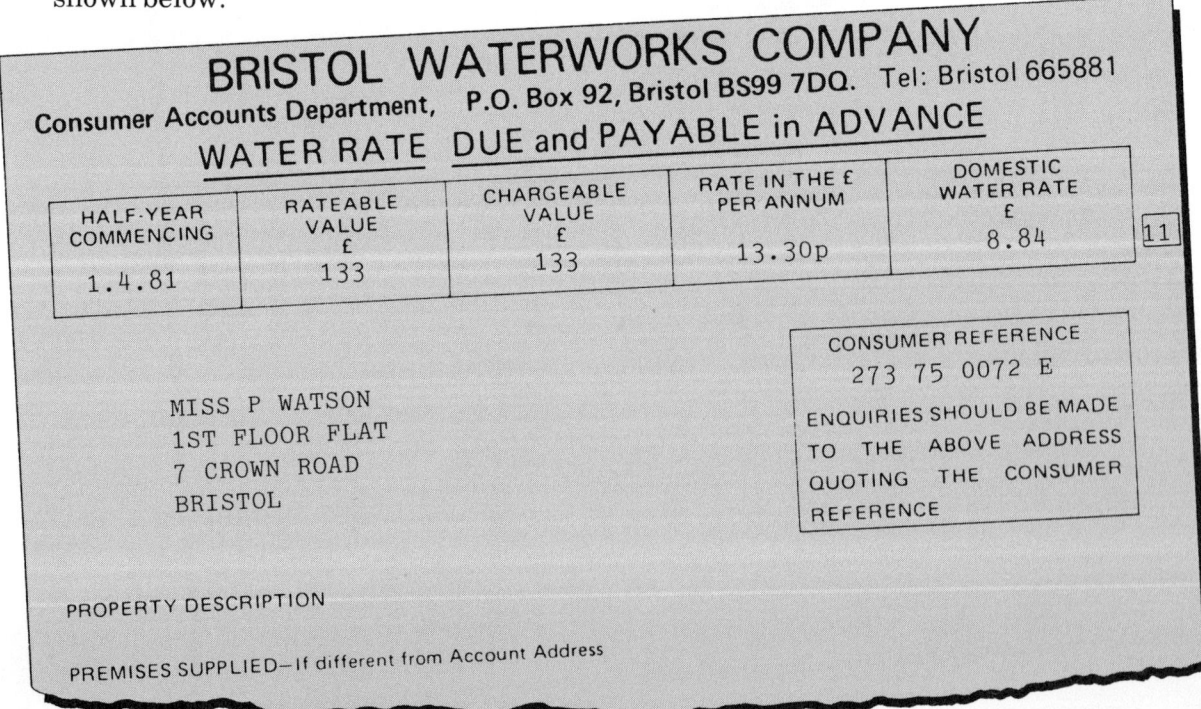

BRISTOL WATERWORKS COMPANY

Consumer Accounts Department, P.O. Box 92, Bristol BS99 7DQ. Tel: Bristol 665881

WATER RATE DUE and PAYABLE in ADVANCE

HALF-YEAR COMMENCING	RATEABLE VALUE £	CHARGEABLE VALUE £	RATE IN THE £ PER ANNUM	DOMESTIC WATER RATE £	
1.4.81	133	133	13.30p	8.84	11

MISS P WATSON
1ST FLOOR FLAT
7 CROWN ROAD
BRISTOL

CONSUMER REFERENCE
273 75 0072 E

ENQUIRIES SHOULD BE MADE TO THE ABOVE ADDRESS QUOTING THE CONSUMER REFERENCE

PROPERTY DESCRIPTION

PREMISES SUPPLIED—If different from Account Address

12 Do you think that Pat Watson should choose to have a meter fitted?

Extension Work

As you know, Mr. and Mrs. Stern live in number 24 with their two children. Mrs. Stern enjoys gardening and she estimates that she uses the hosepipe in the garden for 15 minutes on 20 evenings during the year. With two young children, Mrs. Stern has to use her front-loading washing machine once every day and her dishwasher twice every day. At bedtime the children share a bath, and Mrs. Stern relaxes in a bath late every evening. Mr. Stern has a shower every morning before breakfast.

 13 Collect a Water Metering Decision Sheet. Find the cost for the Sterns to have metered water. Compare this with the cost of unmetered water (which you found in question 3(a)). Which method do you think they ought to choose?

14 Decide the best system for your home, giving reasons.

15 Work out the way in which the Sterns would have to use water (for example, not bathing so often, not using the dishwasher **twice** every day, or using the washing machine every **other** day) so that they would pay no more with a meter than they do by the unmetered system.

5 TV and Video

Which TV?

Mark and Sue want to buy a colour TV set. Sue would like a set that is at least 20 inches or bigger and Mark would like to have a Teletext set. Look at the price list.

1 What is the cheapest set that Sue would like?

2 What is the cheapest set that both Sue and Mark would like?

3 What is the dearest set that they would both like?

4 Find the difference in price between the cheapest and dearest Teletext sets.

Eventually they decide to buy the Philips 3745 22 in. Teletext television. Now read the details of the 5 year guarantee.

5 How much will it cost Sue and Mark to have the 5 year guarantee on the set that they have chosen?

6 What is the total cost of the television and the guarantee?

 7 Collect a Customer Selection Form and a 5 Star Option Form, and fill these in as if **you** were buying the set and the guarantee for cash.

8 Now read the details of the 'confidential hire purchase facilities'. How much deposit would Mark and Sue have to pay for the television and the guarantee using credit terms?

Extension Work

9 (a) Find the total cost for Mark and Sue to buy a GEC C2286 television and the cheapest possible VHS video recorder, both with a 5 year guarantee.
(b) Use a calculator to work out the deposit for this, and how much they would have to pay each month if they used credit facilities.

10 Find the cheapest way that Mark and Sue can buy a television and a VHS video recorder. The television can be any size but it must be colour and it must have a remote control. The video must also have a remote control. Do not include any guarantee costs.

COMET PRICE LIST

Colour T.V.'s

Cat. No.	Description	COMET PRICE inc. VAT
	20 in Models complete with stand	
5902231	DECCACOLOUR DT1475	239.90
5902249	DECCACOLOUR DT8476 remote ...	269.90
5903229	FERGUSON 37350	259.90
5903245	FERGUSON 37351 remote	294.90
5904021	FIDELITY 2022	199.90
5904055	FIDELITY 2024 remote	219.90
5904568	GRUNDIG 6110/6100	229.90
5904584	GRUNDIG 6410 remote	264.90
5905069	ITT CT2500	259.89
5907126	PHILIPS 2026	259.90
5907176	PHILIPS 2226 remote	294.90
5907930	SOLAVOX 20S09	232.90
5907948	SOLAVOX 20R09 remote	259.90
5908164	SONY KV2060 inc. Sony 5 Year Guarantee	298.90
5908106	SONY KV2062 remote, inc. Sony 5 Year Guarantee	349.90
	EXTEND YOUR GUARANTEE TO 5 YEARS FOR £34.95	
	22 in Models complete with stand	
6000210	AMSTRAD CTV2200 remote	239.89
6002335	DECCACOLOUR DV1477	269.90
6002343	DECCACOLOUR 8478 remote	314.90
6003501	FERGUSON 37360	294.90
6003519	FERGUSON 37361 remote	337.90
6004044	FIDELITY 2224 remote	239.90
6004256	GEC C2285	289.89
6004264	GEC C2286 remote	319.89
6005668	GRUNDIG 7100/7110	259.90
6005529	GRUNDIG 7410 remote	309.90
6005430	GRUNDIG 7500 Stereo, 4 speakers, twin channels, 2 x 6 watts, remote	364.90
6006321	ITT 2600	294.89
6006363	ITT 2612 remote	329.89
6006339	ITT CT3493 Stereo, 4 speakers, twin channel, 2 x 10 watts, remote ...	369.90
6007644	PHILIPS 3040	292.90
6007652	PHILIPS 3245 remote	346.90
6008852	SOLAVOX 22S09	259.90
6008860	SOLAVOX 22R09 remote	299.90
6009214	SONY KV2212 remote, inc. Sony 5 Year Guarantee	439.90
	EXTEND YOUR GUARANTEE TO 5 YEARS FOR £34.95	
	26 in Models complete with stand	
6103238	DECCACOLOUR DZ9352 remote ...	369.90
6106587	GRUNDIG 8400 remote	384.90
6106595	GRUNDIG B8600 Stereo, 4 speakers, twin channel, 2 x 30 watts, remote	464.90
6109412	SOLAVOX 26R09 remote	389.90
6109446	SOLAVOX 26SR09 stereo remote ..	479.90
6109975	SONY KV2705UB 27 in remote control 10 watts output, inc. Sony 5 Year Guarantee	559.90
	EXTEND YOUR GUARANTEE TO 5 YEARS FOR £34.95	

Teletext Colour T.V.'s

All following TV's are complete with stand except where starred

Cat. No.	Description	COMET PRICE inc. VAT
5801126	*FERGUSON 37023 16 in. FULL REMOTE CONTROL INC. TELETEXT	329.90
5802287/ 4800012	*GRUNDIG 4404 18 in. FULL REMOTE CONTROL INC. VT1001 TELETEXT	339.90
5804182	*PHILIPS GP4616 16 in. FULL REMOTE CONTROL INC. TELETEXT	369.90
5902273	DECCACOLOUR DT 9476 20 in. FULL REMOTE CONTROL INC. TELETEXT	359.90
5904063	FIDELITY 2026 20 in FULL REMOTE CONTROL INC. TELETEXT	259.89
5904576	GRUNDIG 6400 T/T 20 in FULL REMOTE CONTROL INC. TELETEXT	324.90
5907168	PHILIPS 4626 20 in. FULL REMOTE CONTROL INC. TELETEXT	337.90
5907841	PYE 4266 20 in. FULL REMOTE CONTROL INC. TELETEXT	299.89
5907956	SOLAVOX 20T09 20 in. FULL REMOTE CONTROL INC. TELETEXT	319.90
6000236	AMSTRAD CTV2210 22 in. REMOTE CONTROL INC. TELETEXT	319.89
6002369	DECCACOLOUR DV 9478 22 in. FULL REMOTE CONTROL INC. TELETEXT	379.90
6004060	FIDELITY 2226 22 in. FULL REMOTE CONTROL INC. TELETEXT	299.89
6004272	GEC C2287 22 in. FULL REMOTE CONTROL INC. TELETEXT	399.90
6005553	GRUNDIG 7400 T/T 22 in. FULL REMOTE CONTROL INC. TELETEXT	357.90
6005430/ 4800012	GRUNDIG 7500 22 in. STEREO TV 2 x 6 WATTS, 4 SPEAKERS, TWIN CHANNEL INC. VT1001 TELETEXT	434.90
6006339/ 4801068	ITT CT3493 22 in. STEREO, 2 x 10 WATTS, 4 SPEAKERS, TWIN CHANNEL, REMOTE CONTROL, INC. TRFK 20/24 TELETEXT	449.90
6007694	PHILIPS 3745 22 in. FULL REMOTE CONTROL INC. TELETEXT	397.90
6007628	PHILIPS 3850S 22 in. STEREO TV 2 x 6 WATTS, FULL REMOTE CONTROL, 4 SPEAKERS INC. TELETEXT	489.90

Cat. No.	Teletext Colour T.V.'s continued	COMET PRICE inc. VAT
6008878	SOLAVOX 22T09 22 in. FULL REMOTE CONTROL INC. TELETEXT	359.90
6103246	DECCACOLOUR DZ9358 26 in. FULL REMOTE CONTROL INC. TELETEXT	449.90
6106587/ 4800012	GRUNDIG 8400 26 in. FULL REMOTE CONTROL INC. VT1001 TELETEXT	454.90
6106595/ 4800012	GRUNDIG B8600 STEREO TV, 26 in. 2 x 30 WATTS, FULL REMOTE CONTROL, 4 SPEAKERS, TWIN CHANNEL INC. VT1001 TELETEXT	534.90
6108741	PHILIPS 1250 26 in. FULL REMOTE CONTROL INC. TELETEXT	469.90
6108759	PHILIPS 3880S 26 in. STEREO TV 2 x 6 WATTS, FULL REMOTE CONTROL, 4 SPEAKERS INC. TELETEXT	579.90
6108733	PHILIPS 3895 26 in. STEREO TV 2 x 15 WATTS, FULL REMOTE CONTROL AND SUPERTEXT	689.90
6109496	SOLAVOX 26T09 26 in. FULL REMOTE CONTROL INC. TELETEXT	429.90
6109462	SOLAVOX 26ST09 26 in. STEREO TV, FULL REMOTE CONTROL AND TELETEXT	549.90

Aerials erected at discount prices.

EXTEND YOUR GUARANTEE TO 5 YEARS FOR £42.95

Video Recorders

Cat. No.	Description	COMET PRICE inc. VAT
4000191	AKAI VS1 VHS Recorder, 28 day 1 event timer, 8 channel tuner, infra-red remote control, 5X forward and reverse picture search, on screen interactive monitor	449.90
4000133	AKAI VS4 VHS Double Play Recorder, with interactive monitor system, 28 day 4 event programmable timer, up to 8 hours record/playback. Infra-red remote control 5X forward and reverse picture search, microprocessor logic controls and auto rewind	429.89
4000175	AKAI VS8 VHS Double Play Recorder, with interactive monitor system. Up to 8 hours stereo record/playback with Dolby NR. 28 day 8 event timer, 32 channel tuner, full infra-red remote control. Up to 9X forward and reverse picture search with frame by frame advance	599.90
4001414	AMSTRAD VCR7000 VHS Recorder, 14 day 1 event timer, 8 channel tuner, wired remote control, 4X forward and reverse picture search, freeze frame, pause, BNC sockets for video in/out, phono sockets for audio in/out	359.90
4002020	DECCAVIDEO VRH 8400 VHS Recorder, 14 day 1 event timer, 8 channel tuner, infra-red remote control, 9X forward and reverse picture search, freeze frame and frame by frame advance, auto rewind	419.90
4005329	FERGUSON 3V32, VHS Double play recorder, 14 day 8 event programmable timer. Up to 8 hours record/playback in mono or stereo with Dolby noise reduction, 9X forward and reverse picture search, insert edit facility and motor driven front loading	625.89
4007648	ITT VR3605 VHS Recorder, 14 day 1 event timer, 8 channel tuner, 9X forward/reverse picture search, single frame advance, auto rewind and logic controls	409.89
4007672	ITT VR3905 VHS Recorder, 14 day 1 event timer, 8 channel tuner, infra red remote control, 9X forward/reverse picture search, single frame advance, auto rewind and logic controls	449.89
4007614	ITT VR3975 VHS Recorder, 14 day 8 event timer, 16 channel tuner, stereo record and playback with Dolby NR, 9X forward/reverse picture search, infra-red remote control, single frame advance, front loading and auto rewind ...	519.90
4007656	ITT VR3984 VHS Double Play Recorder, 14 day 8 event programmable timer, 12 channel tuner, 8 hours record/playback in mono or stereo with Dolby noise reduction, 9X forward and reverse picture search, infra-red remote control, insert edit facility	599.90
4000824	PHILIPS VR2022 2000 System Recorder, 16 day, 5 event timer, 26 channel tuner with auto search, 7X forward, 5X reverse picture search with "GO TO" system. Freeze frame up to 8 hours recording and playback, auto rewind	259.89

Rent or buy?

Shortly after Mark and Sue had bought their TV and video recorder, they saw this advert about renting TVs and videos.

Assuming that the TV costs £239.90 to buy, Sue worked out how many months it would take to pay for the set in monthly rental payments. On her calculator she found how many payments of £7.95 equalled the buying price of £239.90.

11 What was Sue's answer?

12 How many months would it be if Sue included in the cost the charge for a 5 year guarantee?

Extension Work

13 Investigate the time taken in months, at £7.95 per month, to pay for the cheapest 20 inch set on the price list, and the dearest 22 inch set. Include the cost of a guarantee. Do you think Mark and Sue were wise to buy instead of renting? Why?

14 Carry out a similar investigation to the one in question 13 for the video rental, using the price list and guarantee costs.

6 The Wedding Reception

Matthew Price is going to marry Alison Brown and plans are being made for the reception at a hotel. Alison's father, Mr. Brown, has some information sent to him with details of various menus and the charges. He knows that the whole reception is going to cost a lot of money!

There are many ways in which he can organise the meal. One way is to have a buffet. There are several different buffet menus to choose from. The price list gives the cost **per person** for the different menus. Look at the price list below:

MENU PRICE LIST

BUFFET MENUS

Buffet A	£6.25
Buffet B	£8.75
Buffet C	£8.25
Buffet D	£9.75
Buffet E	£8.50
Buffet F	£9.50
Buffet G	£11.50
Buffet H	£17.95

MENU SELECTOR

STARTERS

	Price
S1	£1.50
S2	£1.50
S3	90p
S4	£1.10
S5	£2.25
S6	£3.25
S7	M.P.
S8	£2.50
S9	£5.50
S10	£3.95
S11	£2.25

CHEF'S SPECIALITIES

	Price
C1	£10.95
C2	£10.45
C3	£8.50
C4	£8.25
C5	£9.50
C6	£10.95
C7	£8.95
C8	£10.50

SWEET

	Price
D1	£2.00
D2	£1.75
D3	£1.65
D4	£2.50
D5	M.P.
D6	£1.50
D7	£2.50

FISH COURSE

	Price
F1	£3.50
F2	£5.50
F3	£5.50
F4	£5.75
F5	£13.95

	Price
Cheese Course	£1.85
Petits Fours	£1.50
Coffee	95p

All prices are inclusive of VAT.
Chef's Specialities include Vegetables and Potatoes

Mr. Brown has calculated that there will be 76 people at the wedding.

1 Find the total cost for all 76 people for
(**a**) the cheapest
(**b**) the dearest
buffet menus.

Look at the descriptions of the various buffet menus below:

BUFFET MENUS

A
Danish Open Sandwich
Sausage Rolls
Quiche Lorraine
Veal, Ham and Egg Pie
Fried Chicken Legs
Coffee and Cream

B
Cream of Pea Soup
Cold Roast Ham
Cold Chicken
Various Seasonal Salads
Mayonnaise and Vinaigrette Sauce
Baked Jacket Potatoes with Chive Butter
Dutch Apple Pie with Fresh Cream
Coffee and Cream

C
MAINLY HOT FINGER BUFFET
Curried Prawn Bouchees
Hot Chipolatas
Sausage Rolls
Chicken Legs with Devilled Sauce
Quiche Lorraine
Danish Open Sandwiches
Various Gateaux
Board of Cheeses with Celery
Coffee and Cream

D
MAINLY COLD BUFFET
Cold Roast Honey Baked Ham
Roast Wiltshire Turkey
Sirloin of Beef
Hot Dish: Lasagne
Boiled New Parsley Potatoes
Salads: Waldorf, Mixed Rice, Coleslaw,
Italienne, Tossed Green
Profiteroles with Chocolate Sauce
Coffee and Cream

E
MIXED FINGER BUFFET
Seafood Cocktail (Cockles, Prawns)
Assorted Sandwiches:
Ham, Turkey, Beef, Egg
Smoked Salmon with Wholemeal Bread
Veal, Ham and Egg Pie
Quiche Lorraine
Chicken Drumsticks
Scotch Eggs
Fruit Salad and Cream
Belgium Apple Pie and Cream
Coffee and Cream

F
HOT AND COLD FINGER BUFFET
Cheese Souffle Tartlets
Grilled Lamb Cutlets
Goujons of Sole Tartare
Fried Chicken with Barbecue Sauce
Veal and Ham Pie
Pork and Beef Chipolatas
Various Quiches
Curried Seafood Vol au Vents
Cheese and Biscuits
Traditional English Sherry Trifle
Chocolate Gateau
Coffee and Cream

G
Prawn Cocktail
Chicken in Aspic
Raised Pork, Veal and Egg Pie
Roast Rib of Beef
Honey Baked Ham
Salmon Mayonnaise
Salads: Waldorf, Mixed Rice, Coleslaw,
Italienne, Tossed Green
Baked Jacket Potatoes with Chive Butter
Charlotte Russe
Fresh Fruit Salad
Coffee Gateau
Coffee and Cream

H
(minimum of 50 persons)
Melon with Raspberries
Cold Scotch Salmon Mayonnaise
Hot Roast Sirloin of Beef and Mustard
Cold Roast Saddle of Lamb Sauce Vert
Cold Roast Turkey
Cold Honey Baked Ham
Cold Roast Chicken
Beef Stroganoff and Rice
Salads: Waldorf, Mixed Rice, Coleslaw,
Italienne, Tossed Green
Croquembouche
Pineapple Surprise
Cheeseboard and Biscuits
Coffee and Cream

2 (a) Mr. Brown knows that Matthew and Alison prefer beef. Which buffet menus could he choose to give them their favourite type of meat?

(b) How much will it cost for the reception for each of these menus?

3 Which buffet menu would **you** choose?

4 Suppose you were arranging an eighteenth birthday celebration for your friend. There will be 2 parents, your friend and 2 brothers, plus 12 other friends. How much will it cost if you choose your favourite buffet menu?

Another way for Mr. Brown to organise the reception is to have a 'sit-down' meal where the food is served to the guests at their tables. He has to select each course of the meal himself. Now look at the 'Menu Selector'.

5 How much would it cost for each person if Mr. Brown chose a meal of Chilled Melon with Parma Ham, a fish course of Fillet of Sole Bonne Femme, a main course of Escalope of Veal Marsala, and, to finish, Passion Fruit Sorbet, cheese and coffee?

6 What would Mr. Brown have to pay in total if he chose this menu for each of the 76 people at the reception?

 7 Collect a Menu Order Form.

(a) Complete the Buffet Menu Order Form for Mr. Brown if he chose Buffet G.

(b) Complete the Menu Selector Order Form for the reception if Mr. Brown decides to have a sit-down meal, and to:

leave out the fish course and cheese course,

finish, after the dessert, with just coffee, and

choose the cheapest selection in each category.

CONFERENCE & BANQUETING MENU SELECTOR

STARTERS

S.1 Cream of Stilton Soup

S.2 Minestrone Milanaise

S.3 Consommé Celestine

S.4 Cream of Chicken Soup Princess

S.6 Chef's selected Hors D'oeuvres

S.7 Chilled Ogen Melon with Raspberries

S.8 Smoked Fillet of Trout with Horseradish Sauce

S.9 Fresh Poached Salmon Mayonnaise

S.10 Chilled Melon with Parma Ham

S.11 Salmon Mousse Battenburg

FISH COURSES

F.1 **Fillet of Plaice Florentine**
Poached Fillet of Plaice served on a bed of Spinach, coated with a Cream and Cheese sauce

F.2 **Fillet of Sole Bonne Femme**
Poached Fillet of Sole served in a Cream, Parsley, Mushroom and White Wine Sauce

F.5 **Lobster Américaine.**
Half a Lobster served in its shell, coated in a Cream & Brandy Sauce

F.3 **Poached Salmon Hollandaise**
Darne of freshly poached Scotch Salmon served with a Hollandaise Sauce

F.4 **Casserole of Seafood in a Thermidor Sauce.**
Mixed Seafood in a Mustard, Cheese & Wine Sauce served with Saffron Rice

CHEF'S SPECIALITIES

C.1 **Beef Wellington.**
Fillet of Beef coated with a puree of Mushroom and Chicken Liver Paté encased in Brioche Pastry

C.2 **Sirloin of Beef Marchand de Vin**
The finest Sirloin of Scotch Beef served in a rich Red Wine Sauce

C.3 **Supreme of Chicken Neptune.**
Breast of Chicken served with Prawns & Mushrooms in a White Wine Sauce

C.4 **Devilled Poussin.**
Grilled Baby Chicken coated with Mustard and Breadcrumbs and served with a Piquant Sauce

C.5 **Saddle of Lamb Dubarry.**
Roast saddle of Lamb garnished with florets of Cauliflower in a Cheese Sauce and Chateau Potatoes

C.6 **Escalope of Veal Marsala**
Pan-fried Escalopes of Veal served with a Marsala Wine, Cream and Mushroom Sauce

C.7 **Loin of Pork Normandy.**
Roast Loin of Pork served with a Calvados & Apple Sauce

C.8 **Fillet Steak with Pepper Sauce**
(Maximum 100 covers). Pan-fried Fillet Steak served with a Brandy, Green Pepper and Cream Sauce

The above Chef's Specialities include a selection of Fresh Vegetables and Potatoes

DESSERTS

D.1 Charlotte Russe

D.2 Lemon Souffle

D.3 Fruit Flan

D.7 Apple & Walnut Pancake

D.4 Profiteroles with Hot Chocolate Sauce

D.5 Fresh Strawberries (when available)

D.6 Passion Fruit Sorbet

CHEESE

A selection of Local Cheddar, Brie & Stilton

WINES FOR THE TOAST

WHITE WINES

Bin No		Per Bottle
	CHAMPAGNE	£16.50
101	Bollinger NV Special Cuvee	£12.50
102	De LaHaye (Cuvee Privee)	
	WHITE SPARKLING WINES	£ 8.10
110	Pigalle Brut – French NV	

Mr. Brown cannot decide which type of menu to choose for the reception. He is also worried that it is becoming too expensive. After a talk with his bank manager he decides that he ought to limit the amount of money to be spent on the reception to £1000. The bank manager also reminds Mr. Brown that he will need to provide the wine to toast the good health of the bride and groom. Look at the three types of wine available. The price given is for one bottle, and each bottle will serve six people.

8 How many bottles will Mr. Brown need to order for the guests (remember that he has to buy a complete number of bottles)?

9 What is
(a) the cheapest
(b) the dearest
total cost of the wine for the toast?

10 (a) You must now make the final decision for Mr. Brown. Collect a Menu Order Form. Mr. Brown has decided that he will have champagne for the toast but chooses the cheaper type, and that he will have the dearest buffet menu that he can afford. Remember that the total cost must not be over £1000. What will he choose? Complete the order form for him.
(b) If he had the most expensive champagne, which buffet menu would he select then?

Extension Work

11 The groom's father, Mr. Price, telephones Mr. Brown and says that he feels that he ought to contribute something towards the cost of the wedding reception. Mr. Price will pay for the wine (Bin no. 101), and he agrees with Mr. Brown to have Buffet G. Mr. Brown has £275 in the bank. How much must he ask the bank manager to lend him to cover the cost now?

7 Summer Holidays

Paul and Rachel Thomas have two children, Sarah aged 13 and Mark aged 9. They want to go to Ibiza for their summer holidays and they have been looking at the brochures for Ibiza. They decide that they like the town of San Antonio. Look at the descriptions of the hotels in San Antonio.

San Antonio
Ibiza

Hotel Tagomago ★★

With its very good position and high standard of food and service, the Tagomago is an excellent choice of holiday hotel.

The lounge is comfortable and inviting, with a large brick fireplace as its focal point. A friendly bar leads out to an arched terrace, which is scattered with bamboo garden furniture . . . it's a pleasant shady spot for a quiet drink. Not that you need go short of the sun! There's a sparkling swimming pool surrounded by pretty terraces, and the Tagomago is built right at the sea's edge overlooking the bay of San Antonio. It stands on a man-made beach alongside a tiny, natural sandy cove; and you can always catch one of the small boats which call here frequently on their way to other beaches around the bay or to the town centre.

Breakfast, which is served up till 10.45 a.m., consists of a self-service buffet which includes cold meats, cheese and eggs as well as the usual rolls and jams. At dinner the first course is a selection of hors d'oeuvres from which you may help yourself. The main and sweet courses will be served to you at your table.

The bedrooms, like the rest of the hotel, have traditional wooden decor and all rooms have a telephone, bath or shower and toilet. Most also have a terrace, some with partial sea view. The Tagomago is a well established hotel with a pleasantly relaxed atmosphere.
Half board (i.e. dinner, bed and breakfast).

Hotel Tagomago — swimming pool

Hotel Abrat ★★★

The Hotel Abrat stands in large, sunny gardens, which lead down to the rocky shoreline. Just a few yards along the coast you'll find the sandy, man-made beach of Es Calo, where there are facilities for water-skiing and windsurfing, as well as pedalos for hire. If you'd prefer to do your sunbathing in the hotel's very pleasant grounds, you'll find deckchairs and sunbeds scattered over the lawns with trees and bushes providing a welcoming spot of shade here and there . . . it's a totally relaxing scene. The swimming pool and its terraces are equally attractive, and a handy bar out here sells refreshing suntime drinks. There's a tennis court too, plus a paddling pool and a playground for children.

There are several shops and bars within 500 yards, and the lively centre of San Antonio is about 15 minutes' walk away.

The main lounge and bar is friendly and inviting, and there's a second lounge close to the Reception area. There are card tables in one corner of the lounge, and dancing to a band takes place once a week in the lounge/bar. You'll be given a table in the dining room which you'll keep for the length of your stay, and you'll find the service here friendly and efficient.

There's a lift and all bedrooms have a telephone, private bathroom and a terrace. Some also have a full or side sea view.
Half board (i.e. dinner, bed and breakfast).
Full board available.

Hotel Abrat — bedroom

The hotel has lovely gardens

THESE PRICES INCLUDE £11 COVERING AIRPORT TAXES AND PASSENGER CHARGES.

Please read pages 4-17 for general holiday information and page 304 for Booking Conditions.

PRICES PER PERSON IN £'s. HOTEL	ABRAT •								TAGOMAGO •							
Available on Flights	507, 509, 511, 603, 403, 123, 811, 813, 713, 715, 917, 919				509, 811, 713, 917		511, 813, 715, 919		Available on All Flights							
No. of Nights	7		14		10		11		7		14		10		11	
Adult/Child	Ad	Ch	Ad	Ch	Ad	Ch	Ad	Ch	Ad	Ch	Ad	Ch	Ad	Ch	Ad	Ch
13 Apr - 17 Apr / 23 Apr - 4 May	155	Free	196	Free	174	Free	177	Free	151	Free	189	Free	169	Free	171	Free
18 Apr - 22 Apr	170	128	211	128	189	128	190	Free	163	Free	203	Free	182	Free	184	71
5 May - 11 May	167	Free	210	Free	187	74	190	74	164	71	202	74	182	74	184	71
12 May - 18 May	168	74	209	74	215	133	220	133	183	128	233	152	205	140	209	140
19 May - 8 Jun	189	124	246	143	228	160	228	160	187	150	240	190	210	168	215	171
9 Jun - 22 Jun	195	143	257	176	240	190	248	195	198	164	265	215	228	187	235	192
23 Jun - 13 Jul	206	166	283	204	272	230	280	235	221	197	288	249	251	221	258	225
14 Jul - 12 Aug	235	204	317	261	283	237	276	223	235	186	283	237	246	209	253	213
13 Aug - 19 Aug	230	192	312	249	267	218	276	254	208	173	267	218	236	194	241	198
20 Aug - 2 Sep	218	176	281	221	248	198	234	159	200	160	255	202	226	180	231	183
3 Sep - 23 Sep	202	145	259	172	229	158	206	111	183	134	223	165	201	149	204	150
24 Sep - 7 Oct	184	108	225	116	203	113	189	85	166	110	206	141	184	125	187	126
8 Oct - 28 Oct	167	85	208	85	186	85	189	85								

Single room no supplement
Room with terrace 50p
Room with terrace and sea view 70p

SUPPLEMENTS per person per night

Single room £1
Room with sea view or side sea view £1
Full board £1.10

Flights **507, 509, 511** depart from **Gatwick, 603** from **Luton, 403** from **Bristol, 123** from **Cardiff, 809, 811, 813** from **Birmingham, 711, 713, 715** from **East Midlands, 913, 915, 917, 919** from **Manchester, 303** from **Newcastle, 205** from **Glasgow.**

To these prices will be added: 1) Insurance Premiums (£7.60 up to 8 nights, £9.20 up to 17 nights). 2) Individual flight supplements, where applicable, as shown in the Flight Timetable on pages 296-303. For childrens' reductions please see page 10, and for group bookings see page 9.

1 Which of the hotels is best for tennis players?

2 How far from the centre of San Antonio is the hotel Abrat?

3 At the hotel Abrat full board is available. This means that in addition to dinner, bed and breakfast, lunch is provided. Look at the price list to find out how much this lunch would cost.

4 Why is there no supplement in the price list for a room with a terrace at the hotel Abrat?

5 What things do the hotels provide for breakfast?

The Thomas's want to fly from either Bristol or Cardiff airport. They have to decide which two weeks they are going to choose. Paul can have any two weeks off during the period from 22nd June to 22nd July. He finishes work on a Friday at 5 p.m. and he must be able to return to work for a Monday morning at 8.30 a.m. Look at the information below.

6 Write down all the possible departure dates for the Thomas's.

7 Which airport will they have to use?

8 What will the Thomas's flight number be?

June				
Monday	4	11	18	25
Tuesday	5	12	19	26
Wednesday	6	13	20	27
Thursday	7	14	21	28
Friday	1	8	15	22	29
Saturday	2	9	16	23	30
Sunday	3	10	17	24

July					
Monday	2	9	16	23	30
Tuesday	3	10	17	24	31
Wednesday	4	11	18	25
Thursday	5	12	19	26
Friday	6	13	20	27
Saturday	7	14	21	28
Sunday	1	8	15	22	29

Flights From Bristol Airport

Destination Airport (Flying Time)	Flight No.	No. of Nights	British Local Time Dep take off time	Ret. landing time	Departure Dates	Flight Supplements 7 nts	10/11 nts	14 nts
MALLORCA Palma (2h 20m)	401	7/14	Sun 15.25	Sun 21.05	15 Apr - 28 Oct			
IBIZA Ibiza (2h 20m)	403	7/14	Sat 14.50	Sat 13.50	14 Apr - 14 Jul	£19		£19
MENORCA Mahon (2h 15m)	405	7/14	Fri 14.10	Fri 13.10	4 May - 26 Oct	£15		£15
TENERIFE Tenerife Sth (4h 20m)	407	7	Wed 09.25	Wed 18.55	18 Apr	£9		£9
	407	14	Wed 09.25	Wed 18.55	25 Apr, 9,23 May; 6,20 Jun; 4,18 Jul; 1,15,29 Aug; 12,26 Sep; 10,24 Oct	£7		
COSTA DORADA Reus (2h 10m)	409	7	Tue 08.00	Tue 12.55	17 Apr			£7
	409	14	Tue 08.00	Tue 12.55	24 Apr	£3		
	409	14	Tue 08.00	Tue 12.55	8,22 May; 5,19 Jun; 3,17,31 Jul; 14,28 Aug; 11,25 Sep; 9,23 Oct			£3
								£5

Flights From Cardiff Airport

Destination Airport (Flying Time)	Flight No.	No. of Nights	British Local Time Dep take off time	Ret. landing time	Departure Dates	Flight Supplements 7 nts	10/11 nts	14 nts
MALLORCA Palma (2h 15m)	121	7/14	Sun 15.10	Sun 13.20	15 Apr - 15 July			
IBIZA Ibiza (2h 25m)	123	7/14	Sat 14.50	Sat 13.50	21 July - 27 Oct	£19		£19
MENORCA Mahon (2h 15m)	125	14	Mon 15.00	Mon 14.00	16,30 Apr; 14,28 May; 11,25 Jun; 10,24 Sep; 8,22 Oct	£15		£15
	125	7/14	Mon 15.00	Mon 14.00	2 Jul - 27 Aug			£7
ALGARVE Faro (2h 25m)	127	14	Wed 16.15	Wed 15.15	9,23 May; 6,20 Jun; 4,18 Jul; 1,15,29 Aug; 12,26 Sep; 10,24 Oct	£7		£7
								£5

Look at the hotel price list on page 34 again.

9 (a) During which period are the prices the highest for the Hotel Abrat?
  (b) Is the answer the same for the Hotel Tagomago?

10 (a) Write down how much it will cost for each adult for 14 nights at the Hotel Abrat during this period. (See question 9.)
  (b) How much will it cost at the Hotel Tagomago?

The Thomas family decide to stay at the Hotel Abrat. They now have to decide about the rooms. Look again at the hotel price list and you will see a supplement of £1 for a single room at the Hotel Abrat. This means that if only one person is staying in a room then they have to pay an extra £1 per day for the room. Normally the rooms have two beds in them, but it is possible to have an extra bed put in for a child. There is, of course, a reduction for sharing. Look at the table below:

THE FOLLOWING REDUCTIONS (shown as a % discount off the adult price quoted) APPLY TO CHILDREN AGED BETWEEN 2 AND 10 YEARS INCLUSIVE AT THE TIME OF TRAVEL.

		HOTELS			
		1st Child		2nd Child Sharing*	
Departures on or between	Sharing*	Not Sharing*	Sharing with one adult	In hotels showing ●	In all other hotels
6 Apr - 12 Apr	CHILD PRICE IS SHOWN IN THE INDIVIDUAL PRICE PANELS	25%	30%	50%	30%
13 Apr - 22 Apr		5%	10%	15%	10%
23 Apr - 11 May		25%	30%	50%	30%
12 May - 13 Jul		15%	20%	30%	20%
14 Jul - 23 Sep		5%	10%	15%	10%
24 Sep - 28 Oct		25%	30%	50%	30%

* i.e. Sharing with two full-fare paying passengers.

11 (a) How much is the adult price for the Hotel Abrat for a fortnight during the period 23 June–13 July?
  (b) What is the child price?

12 Why will Sarah be charged the adult price?

For the rest of the calculations we have to consider Sarah as an adult.

13 How much cheaper than the adult rate is the child rate for this holiday?

14 (a) What percentage reduction of the adult rate will there be for Mark for this holiday if he does not share with his parents?
  (b) How much money is this?

Paul and Rachel now decide upon several plans for how many rooms they will book and who sleeps in which room. They have drawn up a table to show all of the plans that they think are worth considering. Next they have to calculate the cost for each plan.

Plan	Room 1	Room 2	Room 3
A	Mum + Dad	Sarah	Mark
B	Mum, Dad, Mark	Sarah	—
C	Mum + Dad	Sarah + Mark	—

15 For each of the plans shown above, calculate the total price of the hotel. **Remember:**
   ◆ The holiday is for two weeks during the period 23 June—13 July.
   ◆ Sarah must be considered as an adult. Do not count full board, sea view or flight supplements at this stage.
   ◆ If a room has only one person in it, there is a supplement to pay.
   ◆ **If Mark does not share with an adult, he has to pay the adult price less 15%.**

16 How much will have to be added to any total if the family decides to have full board?

17 Eventually the Thomas's decide that:
   they will choose plan C
   they will have full board
   Mum and Dad's room will have a sea view.
   What will the total cost for this be?

18 We now have to find the total cost of the holiday — you have already found out most of the cost. In addition to the basic price you must make sure that you have taken into account the following:
- ◆ any hotel supplements (see the hotel price list)
- ◆ flight supplements (see the flight information tables)
- ◆ holiday insurance (see the bottom of the hotel price list)

How much is the total cost of the holiday?

> You now have enough information to make a booking.

19 Collect a Holiday Booking Form and complete it for the Thomas's.

20 How much deposit will Paul Thomas have to pay?

21 Paul and Rachel have eight months in which to pay the rest of the money for the holiday. How much must they save each month in order to do this?

Extension Work

The Thomas's will need more money for things like the petrol or train fares to get to the airport, and soft drinks when they are on holiday.

22 Make a list of as many things as possible which the Thomas's may have to pay for in addition to the basic cost of their holiday.

23 Make a reasonable estimate of how much they will need to spend each day while they are on holiday for drinks, ice-cream, snacks, etc. How much will this be for the whole fortnight?

24 How much money should they allow for presents to take home?

25 There are roughly 220 pesetas to the pound (£). How much in pesetas is the full board supplement?

26 Sarah has saved £13.75 for spending on holiday. How many pesetas will Sarah receive when this money is changed into pesetas?

27 If a stamp for a postcard costs 40 pesetas, how much would this be in British money to the nearest penny?

28 Sarah writes a letter to her friend at home. Write a letter that Sarah might send, telling her friend all about the hotel and how she spends her days.

8 Planning a Kitchen

People have to plan a kitchen when they decide to pull out all the units from their old kitchen and start again, or if they have a new kitchen built onto their house as an extension.

Different people like different types of kitchen units. Look at the information given below in the 'kitchens selection chart':

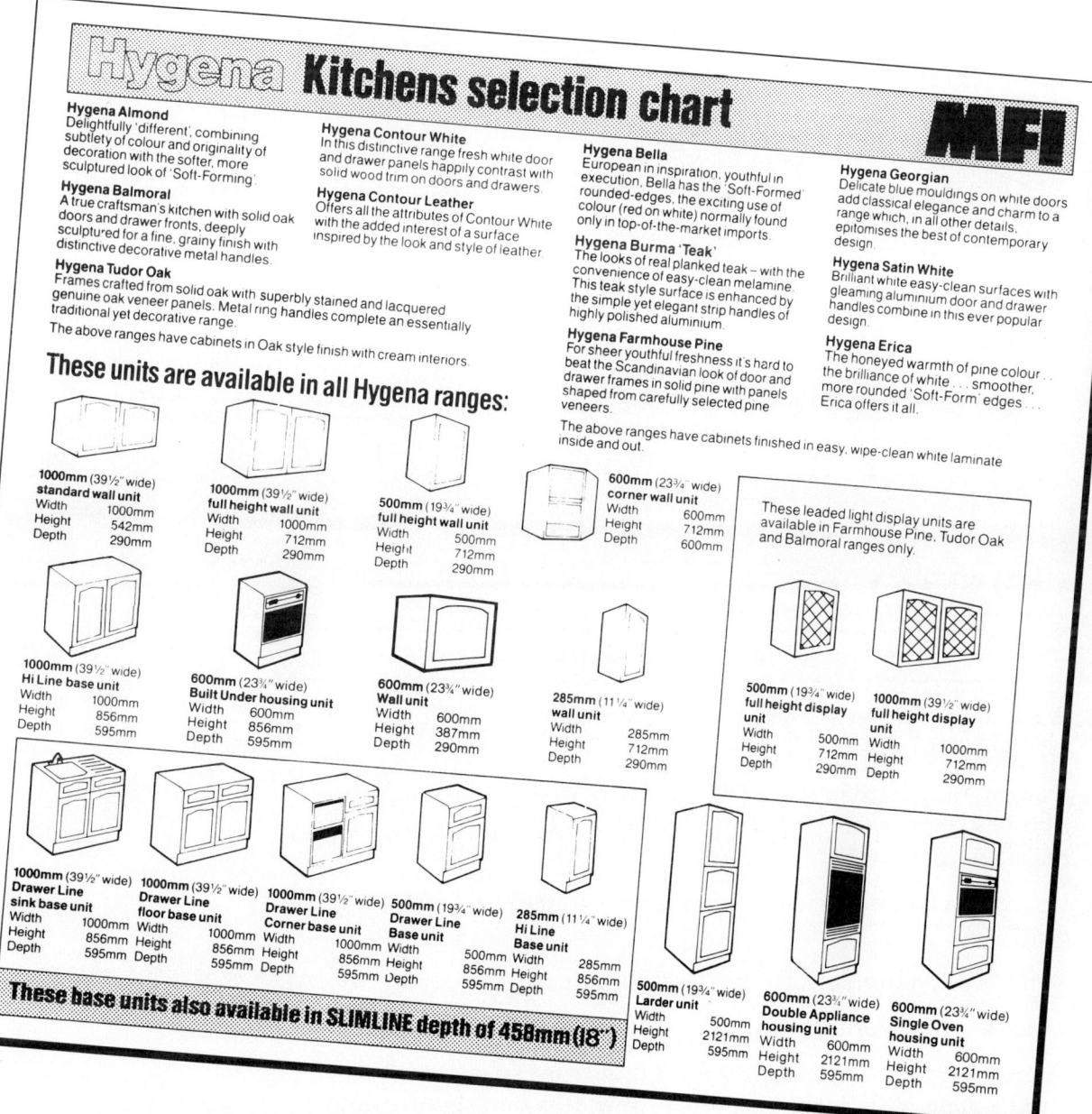

Hygena Kitchens selection chart — MFI

Hygena Almond
Delightfully 'different', combining subtlety of colour and originality of decoration with the softer, more sculptured look of 'Soft-Forming'.

Hygena Balmoral
A true craftsman's kitchen with solid oak doors and drawer fronts, deeply sculptured for a fine, grainy finish with distinctive decorative metal handles.

Hygena Tudor Oak
Frames crafted from solid oak with superbly stained and lacquered genuine oak veneer panels. Metal ring handles complete an essentially traditional yet decorative range.

The above ranges have cabinets in Oak style finish with cream interiors.

Hygena Contour White
In this distinctive range fresh white door and drawer panels happily contrast with solid wood trim on doors and drawers.

Hygena Contour Leather
Offers all the attributes of Contour White with the added interest of a surface inspired by the look and style of leather.

Hygena Bella
European in inspiration, youthful in execution, Bella has the 'Soft-Formed' rounded-edges, the exciting use of colour (red on white) normally found only in top-of-the-market imports.

Hygena Burma 'Teak'
The looks of real planked teak – with the convenience of easy-clean melamine. This teak style surface is enhanced by the simple yet elegant strip handles of highly polished aluminium.

Hygena Farmhouse Pine
For sheer youthful freshness it's hard to beat the Scandinavian look of door and drawer frames in solid pine with panels shaped from carefully selected pine veneers.

The above ranges have cabinets finished in easy, wipe-clean white laminate inside and out.

Hygena Georgian
Delicate blue mouldings on white doors add classical elegance and charm to a range which, in all other details, epitomises the best of contemporary design.

Hygena Satin White
Brilliant white easy-clean surfaces with gleaming aluminium door and drawer handles combine in this ever popular design.

Hygena Erica
The honeyed warmth of pine colour... the brilliance of white... smoother, more rounded 'Soft-Form' edges... Erica offers it all.

These units are available in all Hygena ranges:

1000mm (39½" wide) standard wall unit
Width 1000mm
Height 542mm
Depth 290mm

1000mm (39½" wide) full height wall unit
Width 1000mm
Height 712mm
Depth 290mm

500mm (19¾" wide) full height wall unit
Width 500mm
Height 712mm
Depth 290mm

600mm (23¾" wide) corner wall unit
Width 600mm
Height 712mm
Depth 600mm

These leaded light display units are available in Farmhouse Pine, Tudor Oak and Balmoral ranges only.

1000mm (39½" wide) Hi Line base unit
Width 1000mm
Height 856mm
Depth 595mm

600mm (23¾" wide) Built Under housing unit
Width 600mm
Height 856mm
Depth 595mm

600mm (23¾" wide) Wall unit
Width 600mm
Height 387mm
Depth 290mm

285mm (11¼" wide) wall unit
Width 285mm
Height 712mm
Depth 290mm

500mm (19¾" wide) full height display unit
Width 500mm
Height 712mm
Depth 290mm

1000mm (39½" wide) full height display unit
Width 1000mm
Height 712mm
Depth 290mm

1000mm (39½" wide) Drawer Line sink base unit
Width 1000mm
Height 856mm
Depth 595mm

1000mm (39½" wide) Drawer Line floor base unit
Width 1000mm
Height 856mm
Depth 595mm

1000mm (39½" wide) Drawer Line Corner base unit
Width 1000mm
Height 856mm
Depth 595mm

500mm (19¾" wide) Drawer Line Base unit
Width 500mm
Height 856mm
Depth 595mm

285mm (11¼" wide) Hi Line Base unit
Width 285mm
Height 856mm
Depth 595mm

500mm (19¾" wide) Larder unit
Width 500mm
Height 2121mm
Depth 595mm

600mm (23¾" wide) Double Appliance housing unit
Width 600mm
Height 2121mm
Depth 595mm

600mm (23¾" wide) Single Oven housing unit
Width 600mm
Height 2121mm
Depth 595mm

These base units also available in SLIMLINE depth of 458mm (18")

---

1 Although the 1000 mm standard wall unit only comes in one size, you can buy it with different **finishes** such as Tudor Oak or Farmhouse Pine.
How many different finishes are there for the standard wall unit?

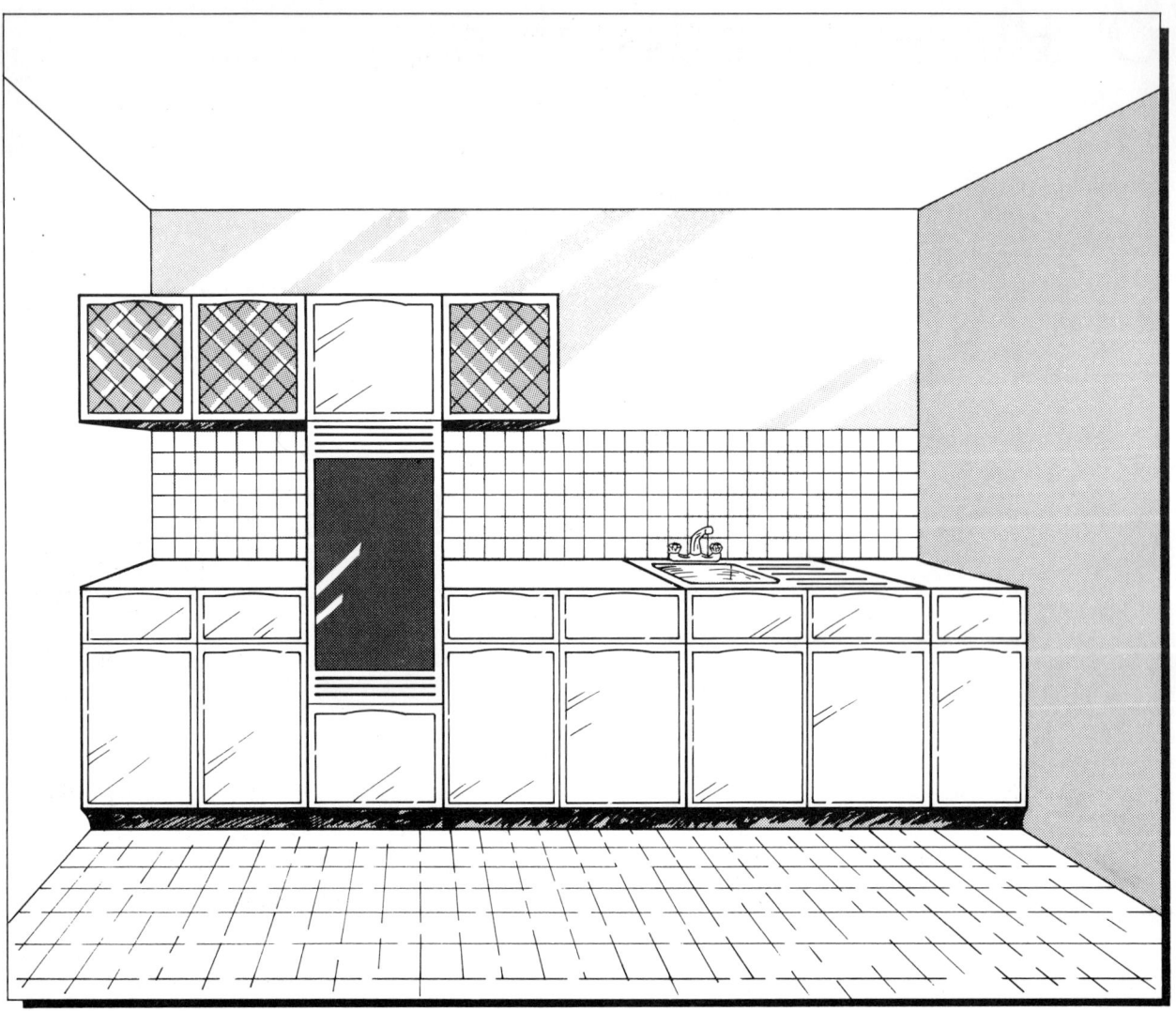

There are 100 cm in 1 m and 10 mm in 1 cm.

2 Write down the measurements of the standard wall unit in:
   (a) cm **(b)** m.

3 One wall in a kitchen is 4 m long, and at one end there is a door that is 858 mm wide. How many 1000 mm full height units can be fitted along this wall?

4 Using measurements in cm, what is the volume of a standard wall unit?

5 What is the volume of a Hi Line base unit, in cm³?

6 There are 3 tall units that stand on the floor. What are they called?

7 There are 6 different base units (5 of them can be bought in slimline depth). What are the different widths that these 6 units come in?

8 Write down the names of all of the units used in the kitchen shown above.

9 If all units are to have the same finish, what finishes could this kitchen be in?

Look at the helpful advice given below. It is intended to help you to make a plan for a kitchen. Next to the advice are some drawings, each showing a kitchen and the plan that goes with it.

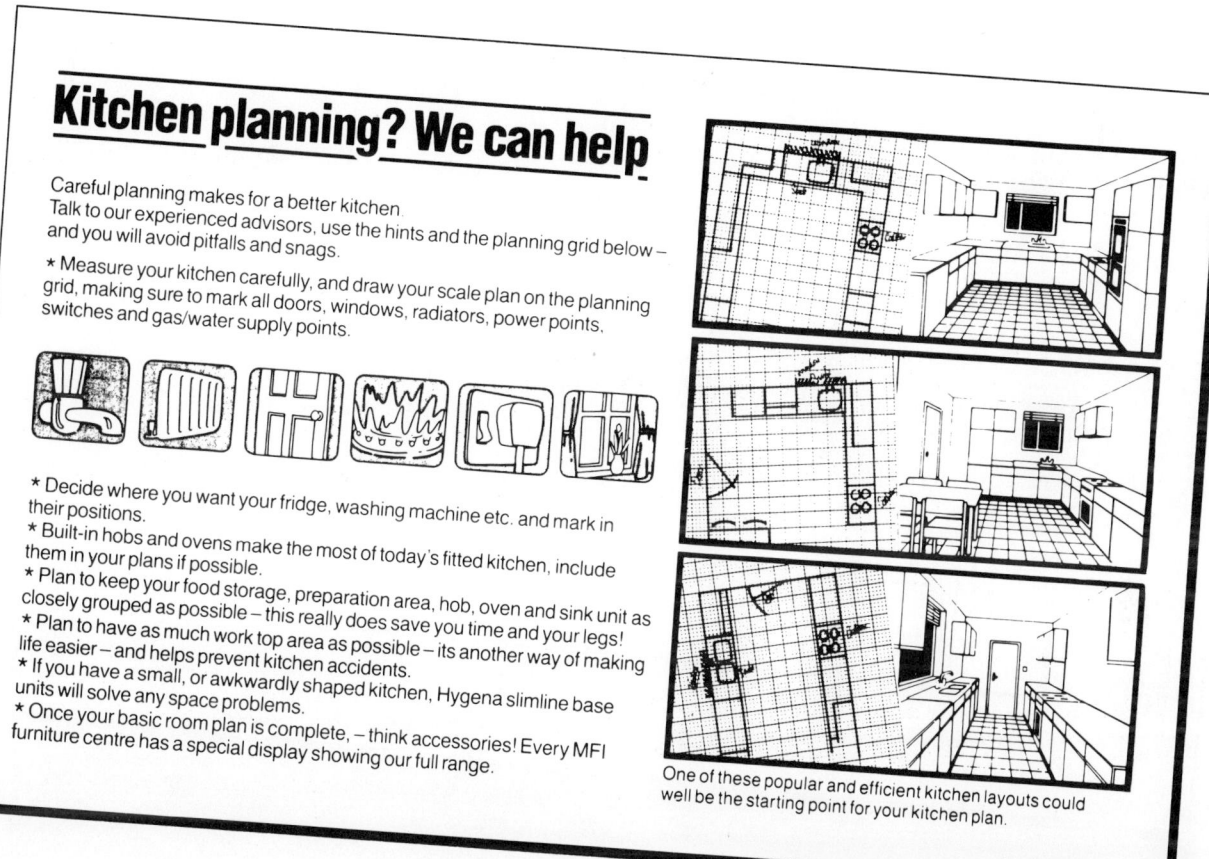

Kitchen planning? We can help

Careful planning makes for a better kitchen.
Talk to our experienced advisors, use the hints and the planning grid below – and you will avoid pitfalls and snags.

* Measure your kitchen carefully, and draw your scale plan on the planning grid, making sure to mark all doors, windows, radiators, power points, switches and gas/water supply points.

* Decide where you want your fridge, washing machine etc. and mark in their positions.
* Built-in hobs and ovens make the most of today's fitted kitchen, include them in your plans if possible.
* Plan to keep your food storage, preparation area, hob, oven and sink unit as closely grouped as possible – this really does save you time and your legs!
* Plan to have as much work top area as possible – its another way of making life easier – and helps prevent kitchen accidents.
* If you have a small, or awkwardly shaped kitchen, Hygena slimline base units will solve any space problems.
* Once your basic room plan is complete, – think accessories! Every MFI furniture centre has a special display showing our full range.

One of these popular and efficient kitchen layouts could well be the starting point for your kitchen plan.

10 Look at the pictures again and see how the plans show a wall unit above a base unit. They are drawn like this because the wall units are not as 'deep' as the base units. Look again at the 'kitchens selection chart'. How far, in mm, will the base units stick out from the wall units?

Look at the plan of a kitchen on the next page. This kitchen has not got any wall units.

11 How long are the windows, in mm?

12 Write down the width and depth of the cooker.

13 (a) Make a list of the different units used in this kitchen.
(b) Write down how many of each you would need to make this kitchen.

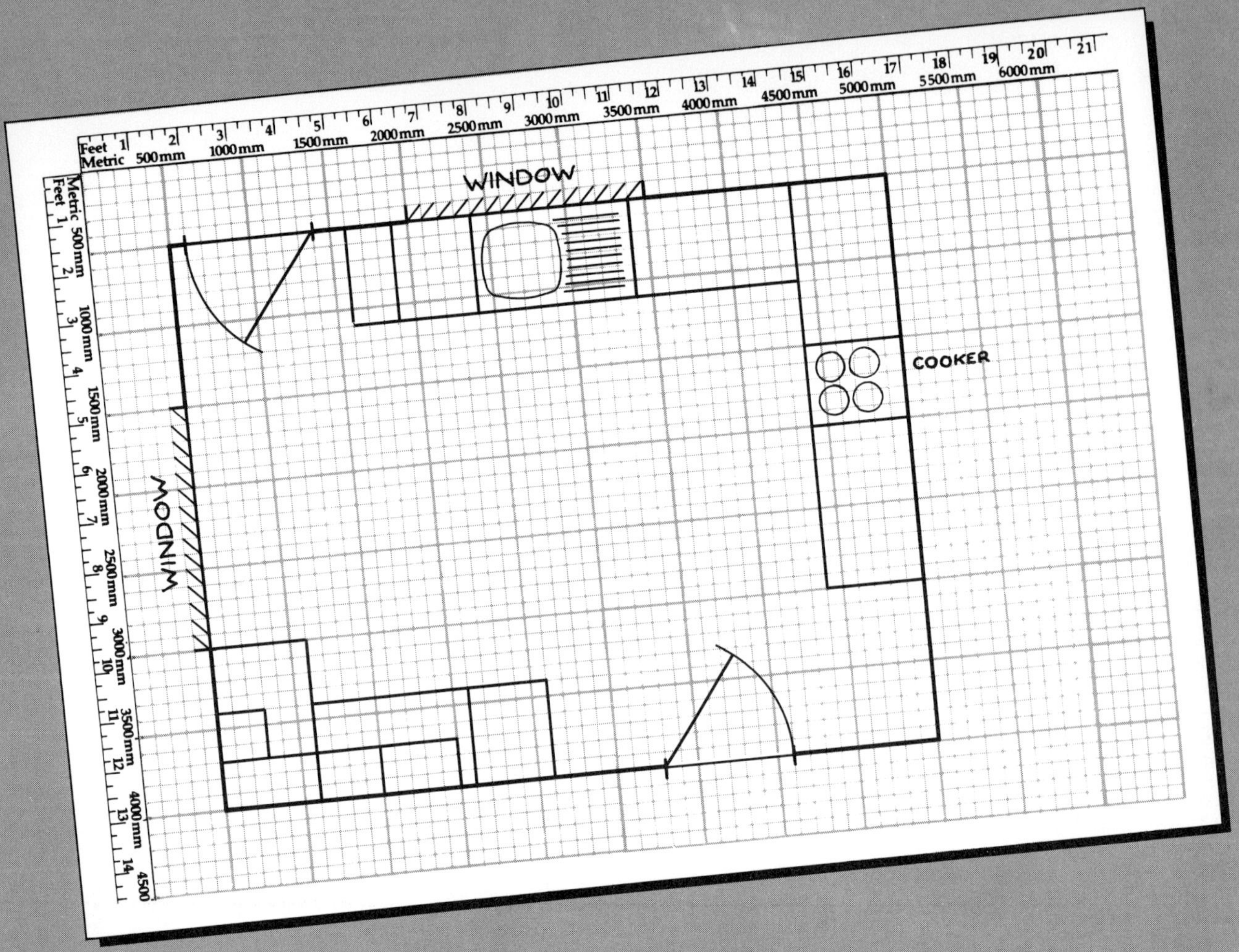

14 (a) Collect a Kitchen Planning Grid and make a copy of the plan on page 42.

(b) Draw on your plan two 1000 mm wall units, one on either side of the cooker.

(c) The wall units in the bottom left-hand corner of the plan are to look like the ones in the drawing on page 42. Draw these on your plan.

15 Why should you not put wall units above the cooker?

 16 Collect a Kitchen Planning: Units Sheet. Cut out the shapes from this sheet for all of the cupboards and wall units used in the kitchen plan you drew in question 14. (You may need to make extra copies of some of the shapes. This can be done by using another sheet, tracing some extra copies, or making your own copies using the squares on the planning grids.) Now fit the cut-out shapes onto your plan.

> Keep your shapes and the rest of the cut-out sheet safe—you will need them to do the rest of the questions in this chapter.

 17 Use a new Kitchen Planning Grid and draw the outline of the kitchen that you drew before with doors and windows. Now use the cut-out shapes to design a **better** kitchen. Use base units and wall units. You must have a sink and a cooker but you could also show where a fridge or a freezer might go. Draw your new design plan on the grid.

18 Write down your reasons for putting certain things in the positions that you have placed them.

19 Explain why you think that this is a better arrangement.

Extension Work

20 Draw a picture of what one wall or a corner of your newly-designed kitchen will look like.

21 Draw a long narrow kitchen and design the arrangement of units.

22 Redesign your own kitchen at home.

23 To fit worktops onto base units can be expensive. Look at the sizes of worktops available on the cut-out sheet you used in question 16.

(a) What tops would you have to order to cover the base units in the kitchen that you drew for question 14?

(b) How much length of worktop would you have to cut off and throw away?

9 Cars

Buying a car can be a very confusing and expensive business. Dave Nash is thinking of buying a Ford Escort. He saw this advertisement in the local newspaper. He does not want a car that was made before 1982.

44

1 How many Escorts are there for sale at Bryans?

2 Describe the cheapest and the dearest Escort models for sale.

Dave goes to the car showrooms and is very interested in the Escort 1.6 GL, priced at £4495. After a long talk with the salesman, Dave asks how much the garage would give him for his car in part exchange. The salesman says £495. In addition, Dave has £1000 saved up.

3 How much will Dave have to borrow in order to afford the Escort 1.6 GL?

In the paper he sees these two adverts:

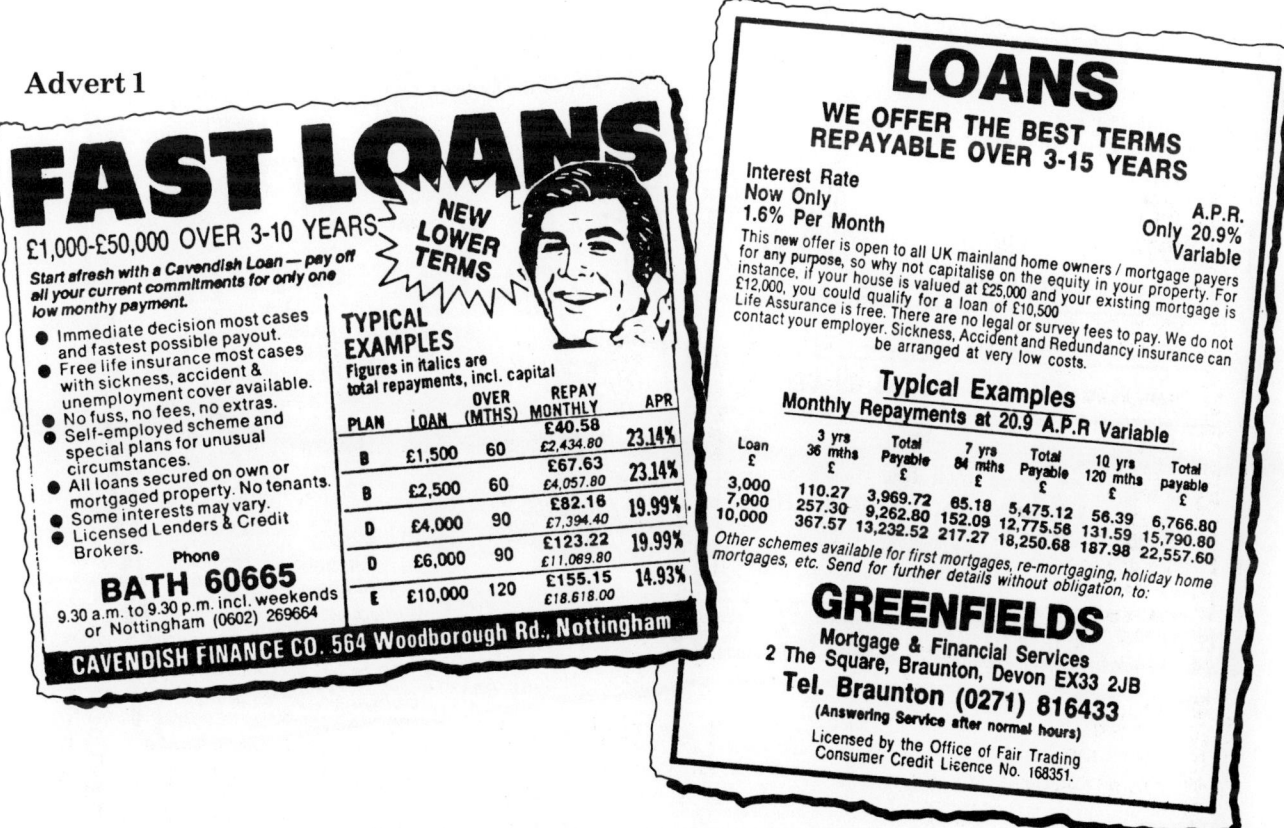

4 How much would Dave have to pay each month if he had a loan from Greenfields over 3 years?

Now look at advert 1. It is not clear how much it would cost to borrow £3000 under plan B, but we can work it out from the figures given. Look at the monthly payments for £2500 and £1500. By subtracting we can find the cost of borrowing £1000, then multiply this by 3 to find the payments for £3000.

5 How much would the monthly payment be to borrow £3000 under plan B from Cavendish Finance Co.?

6 Find out the monthly payment for £3000 borrowed under plan D.

Shortly after buying the Escort, Dave's friend warns him that the two front tyres are nearly bald, and that one of the rear tyres is badly worn. The tyre size is 155 × 13.

Dave sees this advert for tyres, and the voucher giving £1 off each tyre that he buys. He decides that he will have new valves fitted to each tyre and that he will have the front wheels balanced.

7 How much is the difference between the dearest and the cheapest tyre for his car?

8 Dave decides to have Dunlop tyres fitted. How much will he pay altogether for the 3 tyres with the voucher, including valves and balancing?

One day, Dave returns to Bryans at College Green where he bought his Escort just to look around the showroom. In one corner is a brand-new Escort 1.6 GL. He asks the salesman how much it costs, and the salesman shows Dave the new catalogue. The page that Dave is interested in is shown opposite.

9 How much would Dave have to pay, including Car Tax and VAT, for a new version of his car?

10 Work out the final total cost if Dave had the following:

Escort 1.6 GL
Metallic paint
Tinted glass all round
Sunroof

Car Tax is 8.33333% and VAT is 15%. You can check this for the 1.3 GL as follows:

Car Tax = basic price × 8.33333 ÷ 100
= £4942.33 × 8.33333 ÷ 100
= £411.86

Add this to the basic price:
£4942.33 + £411.86 = £5354.19

Now find 15% of this:
VAT = £5354.19 × 15 ÷ 100
= £803.13 (after rounding)

So the total is:
= £4942.33 + £411.86 + £803.13
      basic price   Car Tax   VAT
= £6157.32

11 Do this check for the 1.6 GL.

GL and GL Estate

STANDARD FEATURES

ENGINEERING
- Body:
  Five-door saloon
  Five-door estate
- Engine:
  1.3 OHC (CVH), not estate
  1.6 OHC (CVH)
  1.6 diesel
- Gearbox:
  Four-speed on 1.3
  Five-speed on 1.6 and 1.6 diesel
- Choke:
  Automatic on 1.6
  Manual with integral warning light on 1.3
  Glow-plug pre-warm with warning light on 1.6 diesel
- Tyre size, 155×13

EXTERIOR
- Bodyside moulding:
  Black
  Bright insert
- Bumpers, black with protective inserts
- Door mirrors:
  Passenger side
  Driver side remote control
- Fuel tank cap, locking by ignition key
- Hazard warning flashers
- Heated rear window with auto. switch off
- Handles/locks, black
- Lamps:
  Halogen headlamps
  Reversing lamps
  Rear fog lamps
- Radiator grille, black louvre
- Servo-assisted brakes
- Spoiler, front integral
- Tailgate push lock release
- Tailgate wash/wipe
- Tow hooks, front and rear
- Wheels, styled
- Window surrounds, bright
- Windscreen, laminated
- Windscreen wash, electric
- Windscreen wipers:
  Two-speed
  Intermittent

INTERIOR
- Carpet:
  Load compartment
  Passenger compartment, colour-keyed non-woven velour
- Centre console
- Cigar lighter, front illuminated
- Clock, quartz analogue
- Cloth door inserts
- Coin box
- Courtesy lights:
  Front
  Load compartment
- Fascia instrument switches, illuminated
- Gas struts on tailgate
- Glove box with lid
- Headlining, perforated
- Head restraints on front seats, fully adjustable
- Heater:
  Fan, three-speed
  Illuminated controls
- In-car entertainment:
  Four speakers
  'Joystick' speaker balance control
  Push-button radio (MW/LW)/stereo cassette RST 21P
- Package tray, rear carpeted tilting/removable, saloon only
- Rear view mirror, dipping
- Seats:
  Fabric trim, 'Strobe'
  Front reclining
  60:40 split rear back rest
- Seat belts, front inertia reel
- Steering wheel, two-spoke soft feel
- Stowage bins in front doors
- Stowage shelves
  Driver lower
  Passenger upper
- Trip recorder
- Vanity mirror on passenger sunvisor
- Warning lights:
  Brake failure/handbrake
  'Clean hands'
  Direction indicator
  Ignition/alternator
  Main beam
  Oil pressure

OPTIONAL FEATURES
Factory fitted

Gearbox:	
Five-speed on 1.3	£148.25*
Automatic transmission with oil cooling on 1.6	£433.83*
Central locking with torch key, five-door saloon	£217.45*
Paint, black	£94.35*
Paint, metallic	£109.34*
Tinted glass all round	£50.37*
In-car entertainment	
Push-button stereo radio (MW/LW/VHF)/stereo cassette SRT 32P	£56.21*
Push-button all electronic multi-feature stereo radio (MW/LW/VHF)/stereo cassette ECU1	£193.61*
Seat belts, two rear lap/diagonal inertia and one lap centre static	£105.03*
Sunroof, tilting/sliding glass on saloon only	£302.83*
Windows, electrically operated front on saloon only	£188.71*
Extra cover	see page 126

* Includes Car Tax and VAT

COLOUR AND MAIN PATTERNED TRIM

SOLID COLOURS	'Strobe'
Diamond White	Navy or Steel Grey
Cameo Beige	Mocha or Steel Grey
Rosso Red	Mocha or Steel Grey
Lacquer Red	Mocha or Steel Grey
Cedar Green	Mocha or Steel Grey
Ceramic Blue	Navy or Steel Grey
Ocean Blue	Navy or Steel Grey
Black (at extra cost)	Mocha or Steel Grey

METALLIC COLOURS (at extra cost)	
Strato Silver	Navy or Steel Grey
Quartz Gold	Mocha or Steel Grey
Nimbus Grey	Navy or Steel Grey
Jade Green	Mocha or Steel Grey
Paris Blue	Navy or Steel Grey
Mineral Blue	Navy or Steel Grey
Imperial Red	Mocha or Steel Grey

PRICE GUIDE

Model		Price £	Car Tax £	VAT £	TOTAL £
1.3 GL	5-door	4942.33	411.86	803.13	6157.32*
1.6 GL	5-door	5205.60	433.80	845.91	6485.31*
1.6 diesel GL	5-door	5481.83	456.82	890.80	6829.45*
1.6 GL Estate	5-door	5543.46	461.96	900.81	6906.23*
1.6 diesel GL Estate	5-door	5819.69	484.97	945.70	7250.36*

In order that the salesman can complete an order form for Dave for a new 1.6 GL, he must calculate the basic price of the extras. For example:

The optional extra of black paint has a **total price** of **£94.35**. To take off the VAT, we must divide by 115 and multiply by 100:
$$£94.35 \div 115 \times 100 = £82.04 \text{ (after rounding)}$$
To take off the Car Tax, we divide by 108.33333 and multiply by 100:
$$£82.04 \div 108.33333 \times 100 = £75.73$$
This is the **basic price** of having black paint as an optional extra.

12 Find the basic price of central locking.

13 Find the basic price of the extras given in question 10.

Dave decides to order a new Escort 1.6 GL with the extras given in question 10. He decides to choose the colour Quartz Gold with a Steel Grey trim. To fill in an order form, the salesman needs to know that:

◆ the Vehicle Excise Licence is £100 (no VAT or Car Tax is paid on this)
◆ numberplates have to be fitted, at a cost of £17.85 (before Car Tax and VAT are added)
◆ there is a delivery charge of £105.99 plus VAT of £15.90, making a total of £121.89 (no Car Tax).

 14 Collect a Car Order Form and fill it in to order the car for Dave. Be careful to enter the basic price where you have to, and the Car Tax. Add Value Added Tax as a total at the end for all of the items that have VAT added.

Extension Work

15 If the salesman offers Dave £4200 for his present car, how much extra money must Dave find?

16 **Round this total down to the nearest £500. How much would the monthly payments be if Dave borrowed the money from Cavendish Finance over 5 years?**

17 Dave can only afford to use 25% of his earnings for paying off his loans.
   (a) How much will he owe altogether with this loan **and** his loan to buy the first Escort?
   (b) How much must he be earning to afford these monthly payments if he has one big loan from Cavendish Finance?

18 Look in a brochure to find the cost of the car you'd most like to own.